Resilient and Friendly Living Environments

The paramount goal of this series is to be a groundbreaking resource for professionals, practitioners, and enthusiasts dedicated to shaping resilient, friendly, and inclusive environments. This series aims to empower its readers with cutting-edge insights and practical tools to design spaces that seamlessly blend sustainability, adaptability, and cultural sensitivity. By delving into such topics, the series strives to provide a holistic perspective on creating environments that prioritize the well-being and interconnectedness of diverse communities. Moreover, the series also seeks to guide the development of frameworks that evolve with the ever-changing needs of societies, ensuring that the built environment remains responsive and relevant over time. Ultimately, this book series aspires to be a catalyst for positive transformation, encouraging a paradigm shift in the way we conceptualize, design, and interact with the spaces we inhabit. This series covers a broad spectrum of topics related to building environments that stand resilient and friendly, with a special emphasis on inclusivity for individuals of all ages. The series encompasses but is not limited to:

- Sustainable and Adaptive Design: Guidance on creating environments that are environmentally sustainable and capable of adapting to changing circumstances.
- Cultural Considerations in Design: Insight into designing spaces that respect and reflect the cultural diversity of communities which promote a sense of belonging among individuals.
- Supportive Infrastructure: Detailed examinations of infrastructure elements that contribute to the overall well-being, safety, and accessibility of diverse communities. It can also include wellness initiatives that address physical, mental, and emotional well-being.
- Designing for All Ages: Foster an inclusive environment that values diversity. This includes promoting equal opportunities, and ensuring that everyone feels welcome and respected. In-depth discussions on creating infrastructure and spaces that caters to the unique needs of both children and older adults is also of great importance.
- Community Engagement: Practical approaches to involving the community in the design process and ensuring that the built environment aligns with their needs and preferences.
- Safety Measures and Disaster Risk Reduction: Prioritize safety and security to ensure a comfortable and secure environment. This includes physical safety measures as well as fostering a culture of respect and mutual support.
- Technological Integration: Explorations of how technology can be seamlessly integrated to enhance connectivity, flexibility, and inclusivity within the built environment.
- Collaborative Spaces: Design work and learning environments that facilitate collaboration. This can involve open-plan workspaces, collaborative tools, and flexible seating arrangements that encourage interaction and teamwork.
- Adaptive Policies and laws: Establish policies and laws that are adaptable to changing circumstances. This flexibility allows for the accommodation of diverse needs and ensures that policies remain relevant and effective over time.

This book series is a valuable resource for architects, designers, urban planners, educators, and anyone interested in the intersection of design, community, and well-being. The series aims to inspire and guide readers in creating environments that not only withstand challenges but also flourish as vibrant and inclusive spaces for everyone.

Homa Bahmani · Wei Zhang · Yibin Ao

Socio-Natural Disaster Recovery Projects

Lessons from Recovery After Wenchuan and Bam Earthquakes

Springer

Homa Bahmani
Environment and Civil Engineering
Chengdu University of Technology
Chengdu, Sichuan, China

Wei Zhang
Huazhong University of Science
and Technology
Wuhan, Hubei, China

Yibin Ao
Environment and Civil Engineering
Chengdu University of Technology
Chengdu, Sichuan, China

ISSN 3005-1681 ISSN 3005-169X (electronic)
Resilient and Friendly Living Environments
ISBN 978-981-95-0112-0 ISBN 978-981-95-0113-7 (eBook)
https://doi.org/10.1007/978-981-95-0113-7

This Springer imprint is published by the registered company Springer Nature Singapore Pte Ltd.
The registered company address is: 152 Beach Road, #21-01/04 Gateway East, Singapore 189721, Singapore

Preface I

This book is the result of seven years of extensive research, which started with my Ph.D. thesis. Nevertheless, it all started more than seven years ago when I was thirteen years old and first felt the emotional shock of the Bam earthquake in 2003. I still remember the avalanche of sorrowful news and a singular, heart-wrenching image. A dedicated professor who gave her life on her way to Bam, motivated by an unwavering commitment to save survivors—that picture, together with the deluge of tragic news, remained with me for years. That moment incited in me a firm commitment to understanding the intricate relationship between the forces of nature and human resilience—a determination that has influenced every chapter of this book.

Join me as I take you on a tour of the intricate landscape of recovery projects after socio-natural disasters in this book. What began as my doctoral dissertation has grown into a comprehensive research that integrates theoretical understanding with actual assessments of disaster relief efforts after the Wenchuan and Bam earthquakes, two of the most destructive natural disasters in modern history. In this book, I delve into the idea that recovery is much more than a technical process; it's a life-altering event that reforms neighborhoods, shapes public policy, and revitalizes individuals.

This work is structured more like a story than a roadmap of research steps; it tells the story of disaster recovery from several possible perspectives. To unravel the intricate network of elements that lead to a successful recovery, it uses systems thinking, dives into literature reviews, and draws on the knowledge of field surveys and experts. Reading on, you'll find that the chapters build upon one another, progressively showing how the fundamental human values of hope, persistence, and unity among communities are just as important as safer structural solutions for successful recovery initiatives.

The findings shared here are the product of extensive research and first-hand accounts of the challenges and triumphs of communities hit by disasters. From the first shock to the process of rebuilding and planning, and finally, integrating resilient behaviors into everyday life, I have sought to convey the intricacies of recovery. Academics, professionals, and anybody else affected by socio-natural disasters should find common ground with this storytelling approach.

Without the unwavering support of numerous individuals, this book would not have been feasible. My Ph.D. supervisor, Dr. Wei Zhang, has my eternal thanks for all of the insightful criticism and supportive words that have gone into this project. Dr. Yibin Ao allowed me to broaden this study to include the Wenchuan earthquake, and I owe him a great deal of gratitude. Dr. Alireza Fallahi and Dr. Massomeh Kamaleddini, thank you for your invaluable assistance with the Bam survey. The eleven members of the Delphi panel who provided invaluable feedback also deserve my deepest gratitude. I would also like to thank the Integrated Research on Disaster Risk, Young Scientists Program for supporting this research.

Along the way, my family has also been an invaluable support. My husband for constantly listening to me and talking things out, for proofreading over and over, my parents for their efforts in motivating me when I wanted to give up, and my sister, who went to Bam and acted as my eyes and did the surveys even the pandemic was severe in Kerman. Finally, to all the people of Bam and Sichuan: your kindness and resilience have left an indelible mark on this work.

As you turn these pages, my sincere wish is that you will discover both a scholarly discussion about disaster recovery and an inspiring story of how communities have persevered and rebuilt. Let this story serve as an encouragement for more thought, study, and action toward the goal of creating resilient societies.

Dedicated to the suffered people after the Bam earthquake, Iran, 2003 and Wenchuan earthquake, China, 2008

March 10, 2025

Homa Bahmani
Research Fellow (Research Professor)
Chengdu University of Technology
Chengdu, Sichuan, China

Preface II

Human society is frequently afflicted by natural disasters, which bring about severe human casualties and economic losses. Following a disaster, the reconstruction of infrastructure and communities becomes a critical issue. How to comprehensively, rapidly, and with high-quality rebuild and restore the affected areas to normal, or even surpass previous levels, is an important proposition that requires specialized research.

Systems thinking is a significant research methodology. It treats many important and complex issues as a system, breaking them down into subsystems and factors for analysis one by one. System analysis is a research method that breaks down the complex into simple components, progressively addressing many intricate problems. It analyzes problems as organisms, not just considering the issues in a simplistic manner, but rather examining all aspects comprehensively and analyzing the interrelations and mechanisms within.

Dr. Homa Bahmani has employed the method of systems thinking in her research, which is a primary highlight of the study. By categorizing the post-disaster reconstruction issues into subsystems such as society, economy, environment, engineering technology, and project management, and considering specific factors like government, community, management personnel, funding, materials, and equipment, she has thoroughly considered the steps, key points, and elements of success for post-disaster reconstruction. She has established a comprehensive and systematic framework for the management of post-disaster reconstruction projects, which holds significant referential value.

In her research, Dr. Homa Bahmani has also actively utilized a variety of qualitative and quantitative analysis methods, such as the Delphi method, Triangulation method, and fsQCA method. Through qualitative analysis, she has defined the interrelations between elements of post-disaster reconstruction project management; through quantitative analysis, she has refined the key success factors and influence paths. These conclusions provide a reference for how governments, communities, and project managers can effectively fulfill their respective responsibilities and manage post-disaster reconstruction.

Overall, Dr. Homa Bahmani has conducted very impressive research work. Now, she has compiled her research on post-disaster reconstruction project management into a book for publication. I am honored to write the preface for her book and look forward to even richer research achievements from her.

March 26, 2025

Wei Zhang
Associate Professor
Huazhong University of Science and Technology
Wuhan, Hubei, China

Preface III

According to the Global Assessment Report released by the United Nations Office for Disaster Risk Reduction in 2022, the number of large and medium-sized disaster events is expected to reach 560 per year by 2030, about 1.5 per day on average. Among them—primarily earthquakes, landslides, debris flows, ground collapse, ground cracks, and ground subsidence—ranked first in occurrence. This fact shows that geological disasters have become one of the most important natural disasters in the world, which poses a serious threat to human life safety, infrastructure stability and social and economic development. As one of the countries with the most serious geological disasters and the most threatened population in the world, China faces a complex situation of diverse types of geological disasters, wide spatial distribution, and extremely destructive geological disasters, with 12 types and 48 types of common geological disasters. According to statistics, more than 80% of casualties and losses caused by mountain disasters in China are distributed in rural areas, which has become a key factor in unbalanced and inadequate rural development and restricted the comprehensive revitalization of rural areas. The disaster resistance ability of rural built environment determines the degree of impact of geological disasters on local people's production and life and regional development. The built environment refers to the physical surroundings shaped by human activities, influenced by spatial, temporal, and socio-cultural contexts. The research shows that compared with the urban built environment, the rural built environment is more closely dependent on the natural environment, and its ability to cope with and adapt to geological disasters determines the rural economic development and villagers' social well-being.

Therefore, since 2017, I have been engaged in long-term research on the relationship between the rural built environment and villagers' daily activities and behaviors. In May 2022, Dr. Homa Bahmani joined the team, and we were also honored to participate in the follow-up study of her doctoral thesis research topic. Taking the 2008 Wenchuan Earthquake as a case study, Chapter 7 discusses the key factors contributing to successful post-disaster reconstruction, which hold both theoretical significance and practical value.

During the whole research process, we also received strong support from the doctoral and master's students of the research team and conducted field investigations in the Wenchuan earthquake-hit area. I would like to thank all the members of the team sincerely. Finally, this research is funded by the Major Program of the National Social Science Foundation of China (No. 24 & ZD164).

March 27, 2025

Yibin Ao
Professor
Chengdu University of Technology
Chengdu, Sichuan, China

Competing Interests The authors have no competing interests to declare that are relevant to the content of this manuscript.

Contents

Abbreviations

CIP	Critical Infrastructures Protection
CRED	The Center for Research on the Epidemiology of Disasters
CSFs	Critical Success Factors
fsQCA	Fuzzy-set Qualitative Comparative Analysis
GDP	Gross Domestic Product
HFRI	The Housing Foundation of the Islamic Republic of Iran
IFRCCCS	The International Federation of Red Cross and Red Crescent Societies
IPMI	International Project Management Institute
KEO	Kerman Engineering Office
KPI	Key Performance Index
MC	Mother Consultancy
NCA	Narrative Comparative Analysis
NGOs	Non-Governmental Organizations
PSFs	Primary Success Factors
QCA	Qualitative Comparative Analysis
SFR	Special Fund and Resources
SNDR	Socio-Natural Disaster Recovery
SSFs	Selected Success Factors
SWOT	Strengths, Weaknesses, Opportunities, and Threats analysis
TDDRAI	Top-Down Disaster Recovery Assessment
UNISDR	The UN Office for Disaster Risk Reduction

Chapter 1
Disasters and Current Research Status

Abstract The recovery process after socio-natural disasters is complicated since they are influenced by both natural and human elements. Traditional management practices frequently fail to consider long-term implications, contributing to the high failure rate of disaster recovery efforts. This chapter provides a comprehensive examination of disasters and the available studies in the field of disaster risk reduction by presenting a comprehensive framework that incorporates aspects such as pre- and post-disaster stages, stakeholder viewpoints, and system thinking. Dissecting critical success variables and how they interact with one another provides a methodical framework for better and longer-lasting disaster recovery, closing the gap between theory and practice.

Keywords Disaster recovery · Socio-natural disasters · Project management · Stakeholder engagement · System thinking · Key success factors · Sustainability · Resilience

1.1 Observed Problem and Research Scope

According to The Center for Research on the Epidemiology of Disasters (CRED), a disaster is "*a situation or event that overwhelms local capacity, necessitating a request at the national or international level for external assistance; an unforeseen and often sudden event that causes great damage, destruction and human suffering*" (The Centre for Research on the Epidemiology of Disasters (CRED), 2018). Natural disasters such as earthquakes, floods, and drought that strike without warning can have an impact on human physical (injuries, deaths) or mental (psychological issues) well-being, as well as infrastructure and service loss, economic loss, environmental, organizational, and cultural impacts, among other things (Di Gregorio & Soares, 2017; Dikmen et al., 2012; Dilley et al., 2005). However, various studies have determined the possibility of offering other terms, such as "natural disasters," which convey negative meaning by blaming nature and ignoring humanity's role in vulnerability creation (Chmutina & von Meding, 2019). Therefore, this study applies the

H. Bahmani et al., *Socio-Natural Disaster Recovery Projects*, Resilient and Friendly Living Environments, https://doi.org/10.1007/978-981-95-0113-7_1

term "socio-natural disasters", which simultaneously differentiates these types of disasters from others and shows the role of human activities leading to transform hazards into catastrophes (Johnson et al., 2016).

In 2018, about half the worldwide population was potentially exposed to socio-natural disasters. Among the affected countries, Asia suffered the most. In Asia, 45% of socio-natural disasters, 80% of deaths, and 76% of people affected have been recorded (The Centre for Research on the Epidemiology of Disasters (CRED), 2018). During the last two decades, although the USA suffered the biggest losses (US$ 945 billion) in the world, China represented a higher number of socio-natural disasters (577 against 482) but with less financial loss (US$ 492 billion)(The Centre for Research on the Epidemiology of Disasters (CRED) and The UN Office for Disaster Risk Reduction (UNISDR), 2017). Earthquakes have been ranked as the deadliest form of socio-natural disasters; however, floods affected the highest number of people (The Centre for Research on the Epidemiology of Disasters (CRED), 2018).

According to recent studies, Socio-Natural Disaster Recovery (SNDR) projects, aiming to provide stable and improved facilities for the areas that suffered after disasters, should not only reconstruct the damaged constructions but also provide an opportunity for the city to function better than before. The feature mentioned above is well-known as "Build Back Better". Furthermore, to have sustainable and resilient cities, the decision-makers can take advantage of a logical, structured, and advanced framework for managing SNDR projects (Lu & Xu, 2015; Moatty, 2017; Opdyke, 2017; Tran, 2015; Vahanvati & Mulligan, 2017). Considering the different requirements of these projects during their life cycle may bring a broader perspective toward the different management attitudes, and ideal guidelines can be established. Additionally, the progress of future SNDR projects can be tracked through the success measurement checklists, which are the main outlooks of the last phase.

Based on the World Bank estimation, disaster recovery projects' failure rate is 50%; however, project management can handle this barrier by significantly increasing the chance of success (Supernaw, 2019). Hence, there is a shortage of studies that comprehensively address these projects' needs. While traditional project management usually focuses on time, cost, and quality management, the management of socio-natural disaster recovery projects should be seen as a more complicated task requiring a comprehensive method to address long-term thoughts (Chang et al., 2010; Wu, 2003).

Most scholars studied specific fields in recovery projects following disasters while neglecting the relationships among the subjects. It is necessary to design pre- and post-disaster planning simultaneously. Recovery projects after a socio-natural disaster should also be seen as a portfolio consisting of heterogeneous but dependent projects varying from physical recovery to cultural and economic development. Hence, there is a shortage of a holistic perspective toward SNDR projects to provide an intuitive description for their management. As a result, to conduct a successful SNDR project, an academic-based model covering all the various parts is a prerequisite. Additionally, it is necessary to state clearly the desired goals of SNDR projects. This definition varies across different beneficiaries' perspectives. Therefore, recovery evaluation is expected to differ from different points of view.

This study attempts to mention the following topics:

(1) The definition of a comprehensively successful SNDR project.
(2) Influential factors on the holistic success of SNDR projects.
(3) Consideration of different stakeholders in recovery process assessment.

1.2 Closer Look at Available Literature

1.2.1 Overview of Socio-Natural Disasters and Recovery Projects

Globally, nearly 4 billion people, or approximately 1/2 the worldwide populace, were potentially exposed to natural disasters in 2018 (The Centre for Research on the Epidemiology of Disasters (CRED), 2018). Regarding the UN World Risk Index report for assessing disaster exposure and cities' coping capacity, small countries/ states are the most vulnerable places to natural hazards (Marto et al., 2018). Additionally, small-scale but high-frequency disasters impact communities more than scale and low periodical occurrence (Tran, 2015). Among common natural hazards, climate change, including floods, droughts, hail, frost, and cyclones, constitutes 75–80% of all-natural hazards. In 2014, The United Nations stated that climate disasters have compensated over 40 billion US dollars worldwide (Hu et al., 2018). In 2015, the impacts of socio-natural disasters showed an extremely cascading rate as their global economic loss was US $66 billion. The high degree of relationship among ecological factors leads to intensified socio-natural disasters and widespread poverty. Alongside, socio-natural disasters worsen poverty and make people more fragile against economic failures (Dunford & Li, 2011).

Recovery after socio-natural disasters can be led through two methods. Constructing the hit area as it was before the disaster or Build Back Better (Moatty, 2017). The latter approach can enhance future design, construction, and implementation of damaged buildings or cities. Subsequently, social, economic, and environmental features will experience improvement, leading to boosted disaster preparedness and resilience (Tran, 2015; Vahanvati & Mulligan, 2017). Some researchers misinterpret the concept of Build Back Better in the form of Build Back Safer by just focusing on technical aspects and neglecting socio-economic and environmental elements (Lu & Xu, 2015). On the other hand, the terms "recovery", "reconstruction", restoration", and "rehabilitation" were used in the literature review interchangeably. However, it is preferred to use the term "recovery", which covers a general perspective toward several demands after natural disasters. Recovery should also be seen as an ongoing process that addresses multiple aspects, from emergency response to reconstruction activities (Wu, 2003).

Different needs, experiences, and knowledge derived from various stakeholders are the main reasons for the complex essence of these projects, which necessitates coordination, cooperation, communication, and integration (Supernaw, 2019; Vahanvati & Mulligan, 2017). Besides their dynamic nature, the complicated core of these projects requires dynamic responses (Ismaila et al., 2014; Sadiqi et al., 2017). To solve the problem of complexity in disaster recovery projects, the literature review has been assessed using two methodologies: the chronological method and system thinking. The simultaneous application of these methods can make sure all the possible looks of these projects have been included. Followingly, a comprehensive overview of the literature review based on those two methods has been rendered.

1.2.2 Research Stages About Socio-Natural Disaster Recovery Projects

Although human beings have faced disasters triggered by natural hazards from the dawn of civilization, disasters still have tremendous effects on people and cities due to the shortage of knowledge and preparedness for coping with these catastrophes. Hence, knowledge of previous disaster recovery projects is valuable, as the experience can provide guidelines for future projects. Figure 1.1 illustrates the most mentioned topics in the literature from 2006 to 2023; Fig. 1.2 summarizes the frequencies of subjects among this study's resources.

(1) **2006–2009: Basic idea evaluation**

Appraising some initial topics within this period has been a fundamental step in disaster recovery research. Some studies addressed the critical role of housing reconstruction, including residents' refusal of new houses (Davidson et al., 2007) and comparisons between different housing reconstruction methods (Lyons, 2009). Other studies assessed hasty policy adoption, its long-term impacts (Ingram et al., 2006), and the correlation between disasters and productivity levels (Hallegatte & Dumas, 2009).

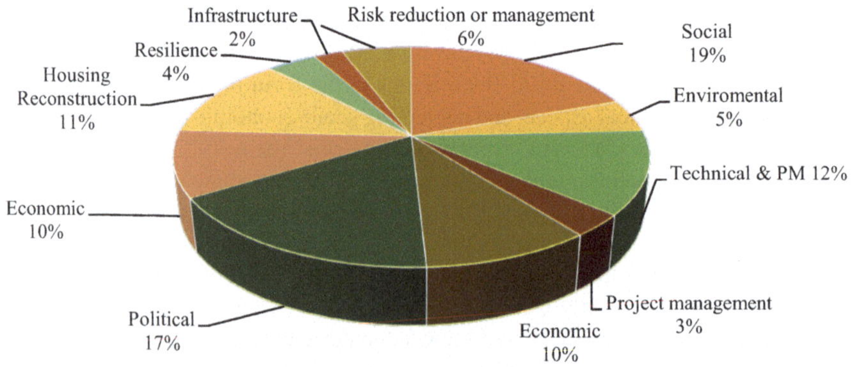

Fig. 1.1 The mentioned topics' frequency within 2006–2023

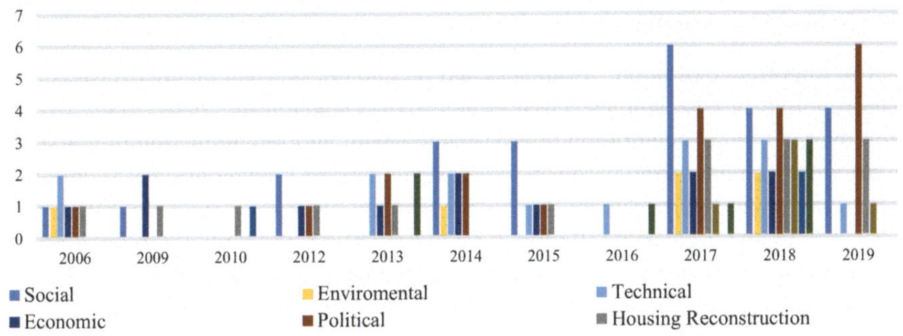

Fig. 1.2 Topics' frequency in recent years

(2) **2010–2013: Subjects' decomposition**

Although housing reconstruction remains a hot research topic, scholars have divided this topic into several separate groups. Johnson and Lizarralde (2012) discussed housing recovery challenges and recommended solutions (Johnson & Lizarralde, 2012). The effects of the technical limitations on project speed, the roles of project management, and challenges were assessed in some studies (Ganapati & Mukherji, 2014). Moreover, studies have examined the impacts of applying the latest technologies on disaster prevention and urban development, such as the seismic spatial information grid (Yang & Zhang, 2013) and cellular automata (Cheng et al., 2013). The approach of the Japanese government towards managing disasters, consisting of pre-disaster plans and post-disaster recovery strategies, was explored in detail (Okamoto & Ishikawa, 2013). In addition, another scholar proposed a resilience-based framework for urban management in disaster recovery projects (Guo, 2012).

(3) **2014–2017: Communication role and project management**

The weaknesses and strengths of previous recovery projects were regarded as the main research topic within this span. A statistical report published by *CRED, 2017* presented a comprehensive analysis of disasters triggered by natural hazards worldwide and an evaluation of their subsequent effects (The Centre for Research on the Epidemiology of Disasters (CRED) and The UN Office for Disaster Risk Reduction (UNISDR), 2017). The differences and similarities between the procedures of disaster recovery projects in different countries have been studied by various researchers (Di Gregorio & Soares, 2017; Xu et al., 2014). Meanwhile, Tran (2015) posed a debatable discussion to assess resilient housing reconstruction in Vietnam (Tran, 2015). Some scholars have discussed the role of society, public participation, and collaboration in these projects (Ismaila et al., 2014; Moatty, 2017; Vahanvati & Mulligan, 2017). During this period, social aspects garnered more attention than other topics. For example, one study appraised social well-being and peoples' livelihoods (Sadiqi et al., 2017), whereas the other research focused on the inevitable roles of stakeholders in building sustainable habitats (Rani et al., 2017).

(4) **2018–2023: Amalgamation of topics**

The importance of public awareness and preparedness when facing disasters has been reported in case studies (Dixit et al., 2018; Okada et al., 2018; Platt, 2018). Moreover, a high concentration of risk management and risk reduction topics such as risk management model based on social participation (Tanwattana, 2018), disaster risk reduction and management models for resilient cities (Etinay et al., 2018), risk prediction in historical areas (Bedini & Bronzini, 2018), and a model for resilience-based management instead of risk-based management (Bostick et al., 2018). Moreover, the researchers' interests have shifted towards the importance of land acquisition (He, 2019; Hosseini et al., 2018). Considering the importance of housing after disasters, a study published in 2019 evaluated the influence of housing materials on the speed of disaster recovery projects through worldwide case studies (Celentano et al., 2019). In addition, the government's role and political decisions in housing these projects have been appraised (Fayazi et al., 2019). Another research showed the relationships between disaster recovery projects and post-conflict reconstruction (Harrowell & Özerdem, 2019). Within the lens of case studies, Li et al. (2019) assessed the housing reconstruction in Dujiangyan after the Wenchuan earthquake to determine the principal reasons for successful housing reconstruction (Li et al., 2019). Moreover, a comprehensive emergency life cycle management model was employed during the Wenchuan earthquake (Han et al., 2019).

1.2.3 Major Aspects of Socio-Natural Disaster Recovery Projects

Considering the study's references, disaster recovery projects have gained more attention in recent years. The study grouped the recently discussed topics into several divisions (Fig. 1.1). This classification was enhanced in Table 1.1, categorizing the topics of the study's primary resources. Chapter three has refined this categorization based on the expanded view gained through evaluating more resources. Subsequently, socio-economic, political, legislation, and technical and project management aspects, the most cited topics in the literature, will be discussed.

(1) **Socio-economic aspects**

Figure 1.3 represents the most mentioned difficulties related to the social subjects, while the suggested solutions have also been proposed. Lack of community capacity, legislation and government attitude, shortage of professional NGOs participating in disaster recovery projects, absence or insufficiency in security, unequal situation for women participation, rigid timeframe in the traditional project management approach, and overemphasizing the technical challenges are common barriers in disaster recovery projects (Sadiqi et al., 2017). The proposed cyclic timeframe by (Okada et al., 2018), which continues beyond the completion of the reconstruction,

Table 1.1 Summary of literature review by applying system thinking

Aspects	Literature	Brief introduction
Socio-economic	(Sadiqi et al., 2017)	Identification of social barriers and proposing objective tree including objectives, purposes, assumptions, and means of verification
	(Okada et al., 2018)	Assessing the consequences of centring communities (Machizukuri) and its effects on people's participation and resilience-based development
	(Sadri et al., 2018)	Evaluating the role of social elements and community ties in the post-disaster recovery process
	(He, 2019)	It suggested a consultative approach following decentralization, a dialogue approach, and an assessment of local residents' needs when people's participation is limited
	(Ganapati & Mukherji, 2014)	Focusing on World Bank's regulations affecting the process and outcome of the housing recovery process and outcomes. The fast development approach applied by the World Bank is contrary to the timeframe of natural disaster recovery projects
	(Hallegatte & Dumas, 2009)	Weather disasters can have positive financial consequences that refer to the "Productivity Effect" or not
	(Marto et al., 2018)	Suggesting an economy-based model to withstand natural disasters, besides fiscal buffers provided by the government for having resilient infrastructures
Political and legislation	(Hinzpeter & Sandholz, 2018)	Emphasizing ecosystem-based solutions for risk reduction strategies. Post Disaster Need Assessment was applied to locate the reasons for environmental ignorance
	(Rani et al., 2017)	Assessing the capability of post-disaster recovery plans to establish resilient and sustainable habitats
	(Dikmen, 2007)	Investigating the reasons for the refusal of newly-built houses by people. The results revealed that the majority of the causes are related to site selection
	(Lu & Xu, 2015)	Finding the most critical Post Disaster Recovery and Reconstruction elements based on real case studies in developed and developing countries
	(Ingram et al., 2006)	Assessing the effects of quick response to post-disaster recovery projects (short-term objectives) on social, environmental, and economic elements (long-term objectives)

(continued)

Table 1.1 (continued)

Aspects	Literature	Brief introduction
	(Lyons, 2009)	Decentralizing housing procurement in post-disaster recovery projects, besides empowerment of poor communities
	(Etinay et al., 2018)	Finding the main principles of Disaster Risk Management regarding Sustainable Development Goals and Disaster Risk Reduction based on the Sendai Framework in resilient urban design
	(Moatty, 2017)	Suggesting flood recovery plans by considering the project's timeframes, stakeholders' roles, territories, and preventive ethics
	(Tanwattana, 2018)	Using the idea of Community-Based Disaster Risk Management in presenting a framework for flood catastrophe management
Political and legislation	(Bostick et al., 2018)	Putting forward the idea of resilience-based management instead of risk-based management
	(Guo, 2012)	Proposing a "Post-Disaster Tourism" model for resilient cities by which economic development can provide an opportunity to promote livelihood
Technical and project management	(Platt, 2018)	Evaluating the impacts of exogenous factors (size of the disaster, demographic, economic) and indigenous factors (authority, decision-making, finance, science, planning) on disaster recovery project's speed and quality
	(Ismaila et al., 2014)	Categorizing Critical Success Factors (CSFs) that contribute to the performance of Non-Governmental Organizations (NGOs) in disaster recovery projects
	(Bedini & Bronzini, 2018)	Considering the community's requirements to suggest programmatic solutions for resurgence based on productive landscape
	(Cheng et al., 2013)	Suggesting a technical process for renewing urban areas and future urban development after a disaster using Cellular Automata
	(Yang & Zhang, 2013)	Presenting a model to manage imminent earthquakes with the use of Spatial Information Grid and seismic information
	(Hidayat & Egbu, 2010)	Exploring the topics of project management in disaster recovery projects through a literature review

(continued)

Table 1.1 (continued)

Aspects	Literature	Brief introduction
	(Blake et al., 2019)	Addressing the factors affecting the efficiency of the transportation system and how resilience can be enhanced in New Zealand
	(Johnson et al., 2016)	Evaluating the relationship among susceptibility, low socio-economic conditions, population density, and natural disaster occurrence
	(The Centre for Research on the Epidemiology of Disasters (CRED) and The UN Office for Disaster Risk Reduction (UNISDR), 2017)	Statistical report of natural hazards, including their categories, occurrence, damage assessment, suffered people's livelihood, and so forth
	(Hosseini et al., 2018)	Putting across a new model for site selection that can be used before a disaster (pre-disaster assessment) based on sustainability factors
Integrated study	(Vahanvati & Mulligan, 2017)	Assessing four case studies following the life cycle, construction phase, and post-construction development. Social, technical, and project management solutions were presented as the final suggestions
	(Di Gregorio & Soares, 2017)	Deducing suggestions for social and technical barriers from American case studies
	(Okamoto & Ishikawa, 2013)	Discussing Japan's experiences in pre-disaster planning and its achievement in post-disaster recovery
	(Tran, 2015)	Defining the lessons learned of Post Disaster Housing Reconstruction focusing on resilient housing reconstruction in different housing methods
	(Johnson & Lizarralde, 2012)	Discussing problematic points in post-disaster housing recovery and making suggestions for more efficient housing reconstruction
	(Dixit et al., 2018)	Inferring critical points about social activities, the role of decision-making, and technical facilities based on Nepal earthquake post-disaster recovery

consists of the pre-disaster phase, post-disaster phase, and maintenance. A comparison study between four disaster recovery projects in Australia and Japan concerned the role of the community and citizen-centred model (Okada et al., 2018). Moreover, disaster recovery projects are affected by several social factors such as people's social networks, living duration in that area (before the disaster), neighbours' assistance,

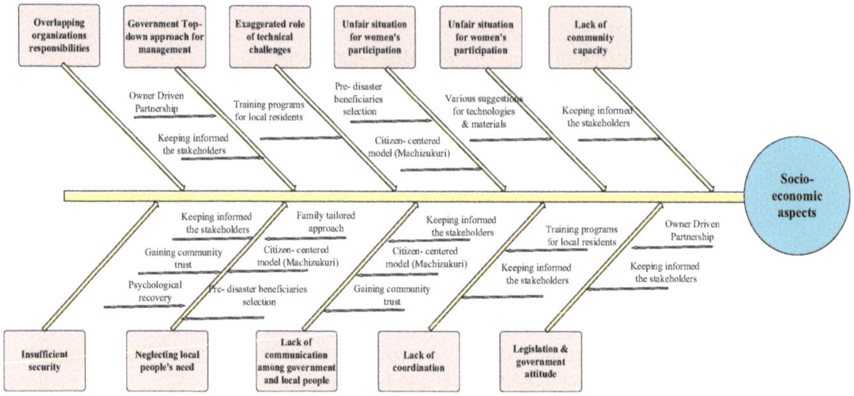

Fig. 1.3 Difficulties and solutions in socio-economic aspects

previous experience facing disasters, and public trust in government (Sadri et al., 2018).

(2) **Political and legislation aspects**

Political (legislation and disaster governing) is published studies' second most widely studied topic. In the research published in 2017, after addressing psychological and social challenges, consideration of government rules and attitudes against temporary and permanent housing have been mentioned as keys for facilitating housing recovery guidelines (Di Gregorio & Soares, 2017). Considering political regulations and bureaucracy, such negligence in policies as ceased policies and plans, concentration on recovery rather than preparation and prevention, vague implementation system for established plans, ambiguous responsibility in bureaucracy, and lack of cooperation between government, civil organizations, and private sectors are common obstacles affecting disaster recovery projects in the political framework (Tanwattana, 2018). Culture dependency and heterogenous perspectives of affected people are also demanding matters that should be solved (Ganapati & Mukherji, 2014). Difficulties and solutions in authority aspects have been graphically shown in Fig. 1.4.

(3) **Technical and project management aspects**

From a project management perspective, speed and quality are some of the most important goals for every project. Indeed, the objectives of some disaster recovery projects are to be done as fast as possible and within qualified outcomes. In a cross-comparison study, fewer casualties in the USA resulted from pre-disaster plans, which had been registered before the earthquake on which building codes, retrofitting masonry buildings, and detailed responsibility breakdown structure were implemented (Lu & Xu, 2015). To reduce technical risks in historical areas in Italy, seismic micro-zoning, light construction structures, and risk theme maps should be applied (Bedini & Bronzini, 2018). As building collapse is regarded as the most

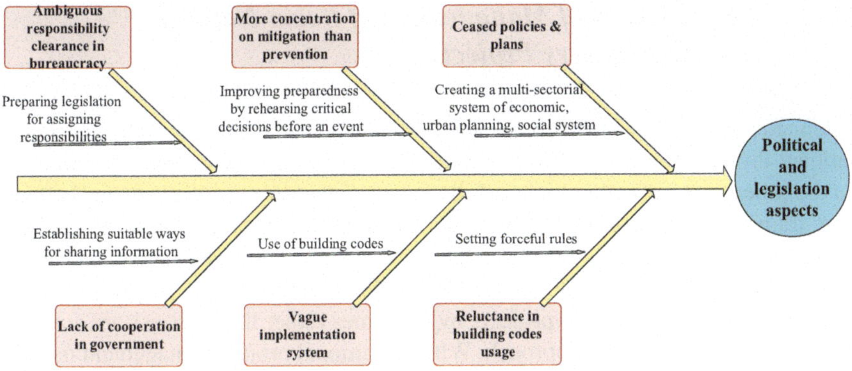

Fig. 1.4 Difficulties and solutions in political and legislation aspects

crucial factor in human casualty, the most vulnerable group of people are the ones who are trapped inside the hit buildings during the earthquake in case of earthquakes. Furthermore, technical and structural improvements for new and existing buildings must be considered. Despite the lessons learned from previous disaster recovery projects, especially in self-built housing reconstruction, people's vulnerabilities related to construction and buildings are being repeated. The public is reluctant to use safe structural elements that hinder the effects of disasters due to financial limitations and the shortage of technical support from technical organizations. Besides, the lack of safety permissions or building permits in governmental rules is another factor in regenerating technical vulnerabilities (Tran, 2015). Figure 1.5 illustrates the difficulties and solutions in technical aspects.

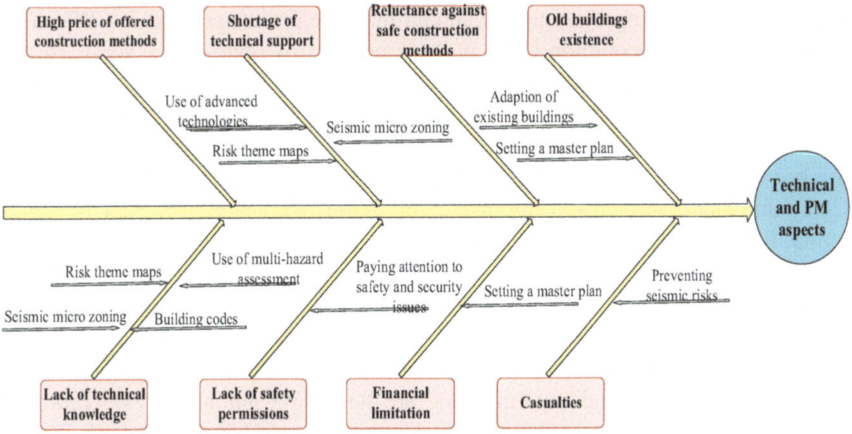

Fig. 1.5 Difficulties and solutions in technical and project management aspects

1.2.4 Special Issues of Management of Socio-Natural Disaster Recovery Projects

A defined project management approach for SNDR projects mentioned by The International Project Management Institute (IPMI) suggests a well-organized framework for the coherent use of technologies, skills, and tools within a specific time and budget. This framework emphasizes different stakeholders' corporations in a closed-loop life cycle. Although the method has advantages, its disadvantages can hinder SNDR projects from being successful; closed-loop (inflexible) time frame, focusing on technical barriers, and evaluating projects by outcomes rather than process are some of the main challenges of this approach. While traditional project management usually focuses on time, cost, and quality management, managing SNDR projects should be seen as a more complicated task requiring a comprehensive method to address long-term (Chang et al., 2010; Wu, 2003). Furthermore, pioneer research presented a project management approach focusing on long-term impacts (Vahanvati & Mulligan, 2017). This section delves into assessing CSFs, barriers, and suggestions in the disaster recovery field from the previous research's points of view.

(1) **Critical Success Factors (CSFs) in SNDR projects**

While the studies have put forward a number of existence cycles and disaster recovery steps, measuring the success of disaster recovery projects needs to consider tangible and intangible outcomes (Akiyama, 2019). These factors can holistically cover the project life cycle and meet different stakeholders' needs. Community participation and network empowerment, effective communication with stakeholders (especially with individuals of the affected groups), considering community culture, and assistance from the neighbourhood authorities are some of the momentous key CSFs (Enshassi et al., 2017; Sadiqi et al., 2017).

One of the most important tools that can measure project success is CSFs. The study listed the most common CSFs in these projects by reviewing several articles regarding different aspects of disaster recovery projects. For a more accurate description, the authors categorized major branches of CFSs. Subsequently, branches have been set for every category based on the literature review. Social and technical factors have been ranked among high-level branches, which consist of more sub-categories. "Community participation and empowerment" is the first ranked CFS from the social factors. Residents are the key stakeholders who suffer from a disaster. These future users of recovery projects should be regarded as the most important part of decision-making, and their needs must be met. After that, among the social branch, "effective communication among stakeholders and managers", "increased public awareness and engagement", "transparency and accountability", "cultural housing design", and "capacity of facilitators and implementors" are the other essential factors. According to the accumulative frequencies of CSFs, as demonstrated by Fig. 1.6, the social and technical factors were often voted on, proving that the numerous technical factors were put forward, notwithstanding the widespread factors from the social sub-systems.

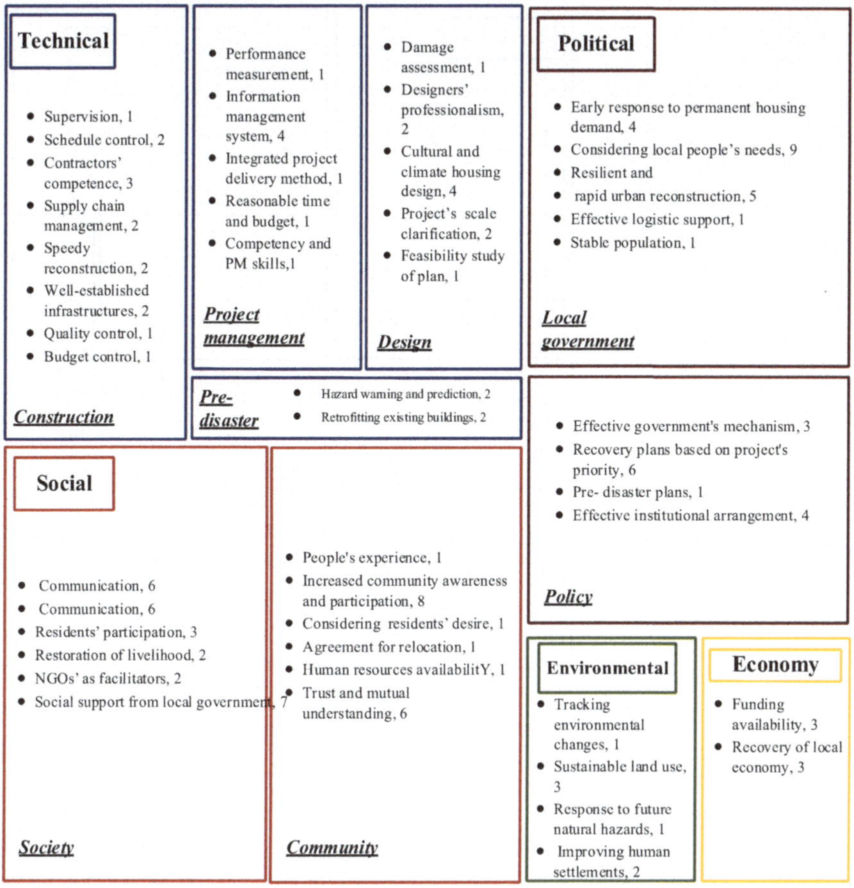

Fig. 1.6 Critical success factors in SNDR projects

(2) **Common barriers in SNDR projects**

The complicated environment of disaster recovery projects necessitates the application of a broader perspective in project management. Despite the different conditions leading to different obstacles in various projects, an in-depth understanding of disaster recovery projects requires discovering prevalent barriers. Furthermore, the authors analyzed to detect the general barriers following the literature review.

The heterogeneous steps in technical barriers generate many factors; however, the political sub-system owns the first-ranked factor, "beneficiaries selection biased toward government" from the policy group (Fig. 1.7). As was revealed within some case studies, the demands of the affected people are ignored because of the government's mechanisms in marginalizing people. In addition, beneficiary information management does not exist, which causes a lack of knowledge of public needs.

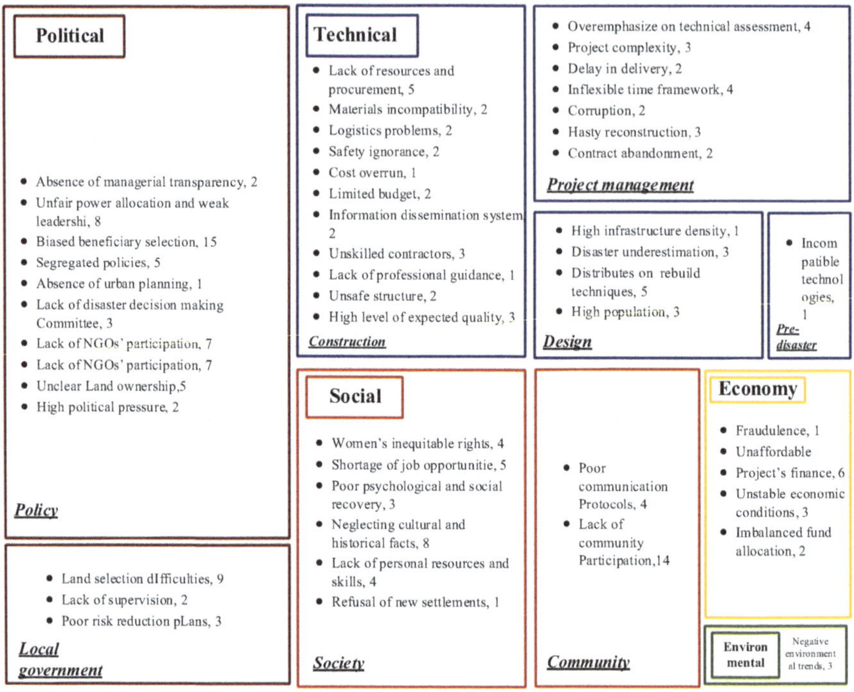

Fig. 1.7 Common barriers in SNDR projects

"Lack of community participation" is the second-ranked factor among the community groups. "Difficulties in land selection for permanent housing", as the third-ranked factor, emphasizes the importance of the land selection process for reconstruction or relocation. This procedure consists of finding a safe place consistent with cultural beliefs to address people's needs properly.

The fourth-ranked factor from the policy group is "unfair power allocation and weak leadership". Unbiased power control within a government or different parties will cause conflicts that hinder a project's success. "Neglecting cultural and historical facts" from the social group as the fourth-ranked factor describes the significance of the designing and planning phase in light of social and historical facts. In the aftermath of the disaster, while the government faces a shortage of resources, NGOs can fulfil a significant role in the recovery process. As a prerequisite for their participation, NGOs' guaranteed safety and stability are addressed as "inappropriate conditions for NGOs' participation". According to Fig. 1.7, the technical topics have several factors whose accumulative frequency is less than the factors from the political sub-system. Although twenty-three barriers are categorized in the technical sub-system, all the highly ranked factors pertain to the political and social sub-systems.

(3) **Proposed suggestions for improving or facilitating SNDR projects**

To alleviate disaster recovery projects' vulnerabilities, logical and empirical solutions must be accompanied by identifying the barriers. Analyzing pre-suggested opinions facilitates proposing comprehensive solutions, which can then work efficiently.

The first-ranked factor, "appropriate operational and evaluation policies" from the policy group, indicates that appraisal procedures and well-established plans significantly contribute to a project's success. According to the second-ranked factor, "effective consultation with stakeholders via communication", the necessity of finding people's needs through communication and participation is stressed. To determine the local public needs and satisfy public demands, sundry solutions are offered based on social-economic and cultural situations. "Owners' participation in housing reconstruction" from society group stands as the next highly ranked factor representing the vital role of residents in self-built houses within the government's framework trying to provide appropriate materials and technologies. Seventy-four factors were classified into technical and social groups, emphasizing researchers' particular concerns regarding construction and social issues. According to Fig. 1.8, the technical group owns the majority of total factors, whereas the higher frequencies of the social factors result in the same accumulation frequencies of these two groups. The political, social, and technical sub-systems have highly ranked factors.

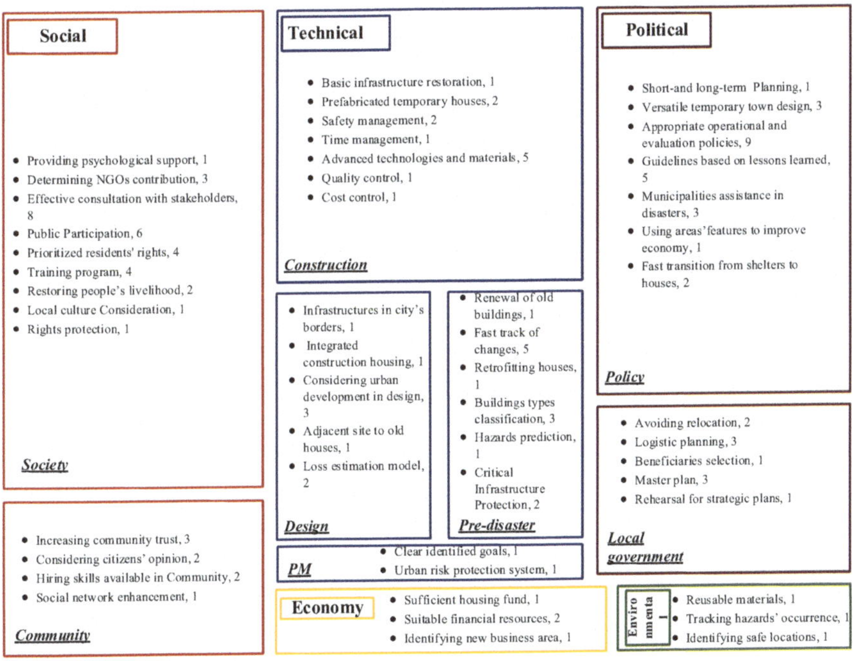

Fig. 1.8 Suggestions for improvement or facilitation of management SNDR projects

1.3 Theoretical Gaps, Research Questions, and Objectives

The literature review manifests some theoretical gaps, which have not found the proper answer within the previous studies. Table 1.2 summarizes the research questions and hypotheses within book chapters.

(1) **Simultaneous consideration of the ongoing life cycle of SNDR project management needs to be enhanced**

Table 1.2 Relationships between literature gaps, research questions, and hypothesis

Research Gaps	Chapters	Research questions	Research hypothesis
Simultaneous consideration of the ongoing life cycle in the management of SNDR projects	Chapter 3	Which pre- and post-disaster steps should be included in SNDR management?	A comprehensive life cycle of SNDR projects might assist in the recognition of the most critical steps in the pre-and post-disaster stages
A well-developed and systematic approach to recognising SNDR projects' goals and the effectual factors	Chapters 3 and 4	Can system thinking assist in identifying the influential factors that influence the success of SNDR projects?	Organizing the scattered information within the SNDR projects' framework by system thinking can create a set of guidelines and illustrate the complexities
Assessing different points of view of SNDR projects among experts and literature	Chapter 4	Which life cycle stages in SNDR projects require further studies to fulfil the real projects' needs?	Some critical aspects might not have gained enough attention in the literature
Comprehensive and time-based success factors, success dimensions, and their interactions	Chapter 5	How do different factors lead to the successful and comprehensive management of SNDR projects?	Success dimensions and affecting factors can be a tracking checklist for better management of SNDR projects
Clarification of the interactions among the factors affecting SNDR projects' success form public perspective	Chapter 6	What combinations of different factors might contribute to successful/ failed recovery per public perspective?	The interactions between the influential success factors may cause a great difference in the success or failure of the project from different stakeholders' points of view
A detailed study on bottom-up and top-down recovery strategies in different contexts and their subsequent outcomes	Chapter 7	Can lessons from the previous recovery projects be helpful for recovery decision-makers?	Different recovery strategies in different countries may lead to various outcomes

As the term recovery may convey the concept of activities happening after a disaster, a disaster recovery project should be seen as an ongoing procedure, starting from the time before the catastrophe and lasting for years after the catastrophe's occurrence. According to the related publications, it is almost accepted that disaster recovery projects have two phases: pre-disaster and post-disaster (Chen & Booth, 2011). While some studies focus on pre-disaster planning and its significant role in preparing the communities to cope with disasters and the negative impacts of lack of preparedness (Xu et al., 2014), some may neglect post-disaster management and planning to reach the resiliency of the suffered area (Wu, 2003). Hence, more effective management of disaster recovery projects requires redesigning the project's steps to cover pre- and post-disaster activities (Jordan, 2012). The disaster management cycle proposed by Patel and Hastak (2013) starts from the emergency phase (Patel & Hastak, 2013). It continues till the forgotten phase, which concentrates on non-structural recovery. This valuable study identified the necessity of simultaneous planning before and after a disaster. However, these projects' proposed cyclic lifespan should be examined, and a practical guideline should be set.

(2) **A well-developed and systematic approach for recognizing SNDR projects' goals and the effectual factors needs to be established**

Although some scholars have studied several disaster recovery project topics, as shown in "integrated studies" in Table 1.1, a systematic framework perceiving the recovery project as a multi-dimensional subject is in its infancy. A recovery project after socio-natural disasters entangles various beneficiaries/ stakeholders whose needs must be appropriately answered to reach a successful project its exact meaning. Some studies validated this fact and suggested considering these projects as complex subjects with various social, structural, and economic issues (Powell, 2011; Wu, 2003). However, plenty of studies focus on particular parts of this complex nature and seek to find successful management pathways for disaster recovery projects. Di Gregorio and Soares (2017) explained that common risks in disasters could be categorized as physical, political, economic, organizational, institutional, cultural, and educational (Di Gregorio & Soares, 2017). One of the earliest studies, which put forward a systematic methodology to see disaster recovery projects, was conducted by Lizarralde (2004). Although this concept was introduced years ago, there is still a shortage of studies to holistically draw a picture of the most significant elements in the success of these projects.

(3) **The critical points of SNDR projects need to be identified based on the opinions of experts**

There is always a difference between what is gained through practice and academic research. While academicians might focus on some critical topics, practitioners and disaster recovery managers might face various challenges in recovery management. Therefore, comparing those groups' ideas is necessary to clarify the shortcomings requiring further research. Moreover, the researcher can validate the qualitative research findings by asking the experts.

(4) **Comprehensive and time-based success factors, success dimensions, and their interactions need to be proposed**

Marginalizing some stakeholders in the projects engaging numerous beneficiaries occurs frequently; however, it can be troublesome in disaster recovery projects. The negligence of beneficiaries due to the government's attitude toward these projects mostly happens in developing countries (Xu et al., 2014). A study conducted in Turkey about the residents' satisfaction with the reconstructed houses evaluated people's low acceptance rate due to the negligence of residents' needs and culture in housing design (Dikmen & Elias-Ozkan, 2016). On the contrary, "Machizukuri", applied in Japan, gained more public satisfaction (Okada et al., 2018).

On the other hand, more focus on technical issues may overshadow the social factors contributing to these projects, which are among the most critical elements to achieving the project's overall success. The necessity of balancing the distribution of resources to answer different aspects' needs should be seen within the project's specific criteria to gain comprehensive success in the recovery process. Furthermore, evaluation of the success of disaster recovery projects should cover both tangible and intangible outcomes (Jordan, 2012).

(5) **The interactions among the factors affecting SNDR projects' success need to be clarified**

CSFs, which are the influential factors in the success of projects, have been discussed. Although recent studies have widely assessed the effectual factors on successful SNDR projects, there is still limited research examining the relationship between CSFs and a specific group of recovery outcomes. While numerous CSFs may affect the recovery outcomes, the influential factors on the recovery success based on the community perspective can be listed as followingly: Effective decision-making and coordination, public participation, public awareness, well-administered and sufficient financial acquisition, resilient rebuilding, available recovery resources (Han, 2014; Ismaila et al., 2014), and consideration of people needs and culture in planning (Moatty, 2017; Sadiqi et al., 2017). Although there is widespread acknowledgement of the positive impact of those factors, disagreement remains about the effect of some parameters. Several studies have focused on maximizing people's participation; however, controlling people's engagement based on local government procedures has offered deliberate public engagement in SNDR projects (Lyons, 2009).

(6) **The recovery strategies in different contexts and their subsequent outcomes in SNDR recovery projects need to be systematically organized**

Disaster recovery does not have fixed pathways, and different communities, even in one country, might show dissimilar recovery results (Jordan, 2012; Li et al., 2016). Research in this area requires examining different countries' experiences in SNDR projects. Several studies have explored the effects of using the disaster recovery lessons learnt. Disaster recovery projects after natural hazards are complex and dynamic projects entangling multiple beneficiaries (Guarnacci, 2012; Opdyke, 2017). Prior works have less focused on the comprehensive comparison of the cases. More recent work evaluating community resilience and vulnerability by comparing socio-natural disaster recovery projects in Japan and Australia illustrated the success of Japan's experience by connecting community members (Okada et al., 2018). Similar ideas proposed in the previous study have been examined through China and Japan's experience coping with the recent socio-natural disasters to access civil society development in those countries (Akiyama, 2019). Although extensive research has been done comparing disaster recovery cases with various characteristics, this area requires further investigation.

Therefore, the research objectives are as follows;

Main objective 1: *Arranging a conceptual framework for integrated management in SNDR projects.*

- Meticulously exploring SNDR projects' life cycle to solve the project stages' difficulties.

Main objective 2: *Providing a holistic set of success dimensions and their influential factors for SNDR projects.*

- Identifying the SNDR project's goals based on stakeholders' views and the project's lifespan.
- Finding out factors associated with the success of the recovery projects following socio-natural disasters.
- Recognition of the combination of the factors may affect the project's success/failure per the specific group of stakeholders.

Based on the research's objectives, governments and legislator organizations can use the research outputs as one of the outcomes is a management framework for successfully and comprehensively managing SNDR projects. Case studies were to present the interactions between the recognized influential factors on the success of the recovery projects at the community level. The research findings can provide a ground for future studies as they aim to simplify the complexity of SNDR projects' management via the application of system thinking.

1.4 Overview of the Book Structure

Chapter 1 systematically and chronologically explores the relevant published literature. The chapter will continue discussing the management of SNDR projects and end by identifying the literature gaps. The research strategy and process, data collection, and data analysis methods, alongside the detailed roadmap of the research, have been presented in Chapter 2. System thinking has divided SNDR projects into five sub-systems in chapter 3. This chapter proposed a conceptual management framework based on continual SNDR projects' life cycle and project management stages. The framework in Chapter 3 is a management tool supposed to be applied in real recovery projects for better performance. Therefore, due to research limitations, the researchers have not applied it in cases. A combination of the suggested life cycle and a systematic categorization of the identified factors (system thinking) have presented the Primary Success Factors (PSFs) list in Chapter 4. Moreover, verifying the PSFs using the Delphi survey generated the Selected Success Factors (SSFs). A critical literature review and comparison of Delphi results have also shed light on future research needs. In Chapter 5, the researchers utilized triangulation of the results of the last methods and Qualitative Comparative Analysis (QCA) to construct Critical Success Factors (CSFs) and Key Performance Indicators (KPIs) in socio-natural disaster recovery projects. The findings in Chapter 5 have been applied in the two following chapters through comparative case studies. Considering people as one of the most critical stakeholders in SNDR projects, public points of view have been considered to examine the success pathways through fuzzy set Qualitative Comparative Analysis (fsQCA). The community's perspective toward successful recovery is then evaluated in Chapter 6. Success/failure pathways of Bam's recovery per community perspective (public semi-structured interviews) are the combinations of the factors assessed. Additionally, correlations among the effectual parameters have been assessed. Chapter 7 holistically assessed the top-down and bottom recovery assessment of two disaster recovery projects in Iran and China following the Bam and Wenchuan earthquakes. We applied mixed methodology to provide unique perspectives from different points of view of stakeholders involved in disaster recovery projects. Finally, Chapter 8 concludes the research findings and significance and puts forward suggestions for future research.

References

Akiyama, H. (2019). *Who rebuilds? Civil society in post-disaster recovery in China and Japan.* George Mason University.

Bedini, M. A., & Bronzini, F. (2018). The post-earthquake experience in Italy. Difficulties and the possibility of planning the resurgence of the territories affected by earthquakes. *Land Use Policy, 78*(November 2018), 303–315. https://doi.org/10.1016/j.landusepol.2018.07.003

Blake, D. M., Stevenson, J., Wotherspoon, L., Ivory, V., & Trotter, M. (2019). The role of data and information exchanges in transport system disaster recovery: A New Zealand case study. *International Journal of Disaster Risk Reduction, Upcoming.* https://doi.org/10.1016/j.ijdrr.2019.101124

Bostick, T. P., Connelly, E. B., Lambert, J. H., & Linkov, I. (2018). Resilience science, policy and investment for civil infrastructure. *Reliability Engineering and System Safety, 175*(July 2018), 19–23. https://doi.org/10.1016/j.ress.2018.02.025

Celentano, G., Escamilla, E. Z., Göswein, V., & Habert, G. (2019). A matter of speed: The impact of material choice in post-disaster reconstruction. *International Journal of Disaster Risk Reduction, 34*(2019), 34–44. https://doi.org/10.1016/j.ijdrr.2018.10.026

Chang, Y., Wilkinson, S., Seville, E., & Potangaroa, R. (2010). Resourcing for a resilient post-disaster reconstruction environment. *International Journal of Disaster Resilience in the Built Environment, 1*(1), 65–83. https://doi.org/10.1108/17595901011026481

Chen, Y., & Booth, D. C. (2011). The Wenchuan earthquake of 2008 anatomy of a disaster. In *Zhurnal Eksperimental'noi i Teoreticheskoi Fiziki.* Springer. https://doi.org/10.1007/978-3-642-21159-1

Cheng, Y., He, Z., Xie, Y., & Deng, H. (2013). Modeling of urban post-disaster reconstruction using extended cellular automata. In F. Bianl & Y. Xie (Eds.), *International Conference on*

Geo-Informatics in Resource Management and Sustainable Ecosystem (pp. 86–97). Springer. https://doi.org/10.1007/978-3-662-45737-5

Chmutina, K., & von Meding, J. (2019). A dilemma of language: "Natural Disasters" in academic literature. *International Journal of Disaster Risk Science, 10*(3), 283–292. https://doi.org/10.1007/s13753-019-00232-2

Davidson, C. H., Johnson, C., Lizarralde, G., Dikmen, N., & Sliwinski, A. (2007). Truths and myths about community participation in post-disaster housing projects. *Habitat International, 31*(2007), 100–115. https://doi.org/10.1016/j.habitatint.2006.08.003

Di Gregorio, L. T., & Soares, C. A. P. (2017). Post-disaster housing recovery guidelines for development countries based on experiences in the American continent. *International Journal of Disaster Risk Reduction, 24*(2017), 340–347. https://doi.org/10.1016/j.ijdrr.2017.06.027

Dikmen, N. (2007). Relocation or rebuilding in the same area: An important factor for decision making for post disaster housing projects. In D. Alexander, C. H. Davidson, A. Fox, G. Lizzaralde & C. Johnson (Eds.), *Proceedings of the International Conference and Student Competition on Post-disaster Reconstruction - Meeting Stakeholder Interests*. Firenze University Press.

Dikmen, N., & Elias-Ozkan, S. T. (2016). Housing after disaster: A post occupancy evaluation of a reconstruction project. *International Journal of Disaster Risk Reduction, 19*, 167–178. https://doi.org/10.1016/j.ijdrr.2016.08.020

Dikmen, N., Elias-Ozkan, S. T., & Davidson, C. (2012). Comparison of post-disaster housing procurement methods in rural areas of Turkey. *Open House International, 37*(1), 28–39.

Dilley, M., Chen, R. S., Deichmann, U., Lerner-Lam, A., Arnold, M., Agwe, J., Buys, P., Kjekstad, O., Lyon, B., & Yetman, G. (2005). Natural disaster hotspots: A global risk analysis. In *World bank disaster risk management series* (Issue 5). The World Bank.

Dixit, A. M., Shrestha, S. N., Guragain, R., Pandey, B. H., Oli, K. S., Adhikari, S. R., Acharya, S. P., Jimee, G. K., Upadhyaya, B. K., Sangachhe, S. B., Shrestha, N., Pradhan, S., Dhungel, R., Khatiwada, P., Baskota, A., Poudel, A., Tandingan, M., Mathema, N. U., Parajuli, B., & Marasini, N. (2018). Risk management, response, relief, recovery, reconstruction, and future disaster risk reduction. In *Impacts and insights of the Gorkha earthquake* (pp. 95–134). Elsevier Inc. https://doi.org/10.1016/B978-0-12-812808-4.00005-5

Dunford, M., & Li, L. (2011). Earthquake reconstruction in Wenchuan: Assessing the state overall plan and addressing the "forgotten phase." In *Applied geography* (Vol. 31, Issue 3, pp. 998–1009). https://doi.org/10.1016/j.apgeog.2011.01.001

Enshassi, A., Chatat, T., von Meding, J., & Forino, G. (2017). Factors influencing post-disaster reconstruction project management for housing provision in the Gaza Strip, occupied Palestinian territories. *International Journal of Disaster Risk Science, 8*(4), 402–414. https://doi.org/10.1007/s13753-017-0155-4

Etinay, N., Egbu, C., & Murray, V. (2018). Building urban resilience for disaster risk management and disaster for risk reduction. *Procedia Engineering, 212*(2018), 575–582. https://doi.org/10.1016/j.proeng.2018.01.074

Fayazi, M., Yeh, E. T., & Li, F. (2019). Development and divergent post-disaster trajectories in a mountain village: Temporal dynamics of differentiation after the 2008 Wenchuan earthquake. *World Development, 124*(2019), Article 104663. https://doi.org/10.1016/j.worlddev.2019.104663

Ganapati, N. E., & Mukherji, A. (2014). Out of sync: World bank funding for housing recovery, postdisaster planning, and participation. *Natural Hazards Review, 15*(1), 58–73. https://doi.org/10.1061/(asce)nh.1527-6996.0000120

Guarnacci, U. (2012). Governance for sustainable reconstruction after disasters: Lessons from Nias. *Environmental Development, 2*(2012), 73–85. https://doi.org/10.1016/j.envdev.2012.03.010

Guo, Y. (2012). Urban resilience in post-disaster reconstruction: Towards a resilient development in Sichuan, China. *International Journal of Disaster Risk Science, 3*(1), 45–55. https://doi.org/10.1007/s13753-012-0006-2

Hallegatte, S., & Dumas, P. (2009). Can natural disasters have positive consequences? Investigating the role of embodied technical change. *Ecological Economics, 68*(3), 777–786. https://doi.org/10.1016/j.ecolecon.2008.06.011

Han, J., Zhang, P., & Song, Y. (2019). The construction of emergency communication model based on the Kite-Type co-orientation approach. In H. Tan (Ed.), *Technology for education* and *learning* (Vol. 53, Issue 9, pp. 235–243). Springer. https://doi.org/10.1017/CBO9781107415324.004

Han, Z. (2014). *From vulnerability to resilience: Long-term livelihood recovery in rural china after the 2008 Wenchuan earthquake*. University of Delaware.

Harrowell, E., & Özerdem, A. (2019). Understanding the dilemmas of integrating post-disaster and post-conflict reconstruction initiatives: Evidence from Nepal, Sri Lanka and Indonesia. *International Journal of Disaster Risk Reduction, 36*(2019), Article 101092. https://doi.org/10.1016/j.ijdrr.2019.101092

He, L. (2019). Identifying local needs for post-disaster recovery in Nepal. *World Development, 118*(2019), 52–62. https://doi.org/10.1016/j.worlddev.2019.02.005

Hidayat, B., & Egbu, C. (2010). A literature review of the role of project management in post-disaster reconstruction. In Procs *26th Annual ARCOM Conference* (pp. 1269–1278), September.

Hinzpeter, K., & Sandholz, S. (2018). Squaring the circle? Integrating environment, infrastructure and risk reduction in post disaster needs assessments. *International Journal of Disaster Risk Reduction, 32*(December 2018), 113–124. https://doi.org/10.1016/j.ijdrr.2018.05.016

Hosseini, S. M. A., Pons, O., & de la Fuente, A. (2018). A combination of the Knapsack algorithm and MIVES for choosing optimal temporary housing site locations: A case study in Tehran. *International Journal of Disaster Risk Reduction, 27*(July 2017), 265–277. https://doi.org/10.1016/j.ijdrr.2017.10.013

Hu, H., Lei, T., Hu, J., Zhang, S., & Kavan, P. (2018). Disaster-mitigating and general innovative responses to climate disasters: Evidence from modern and historical China. *International Journal of* Disaster *Risk Reduction, 28*(June 2018), 664–673. https://doi.org/10.1016/j.ijdrr.2018.01.022

Ingram, J. C., Franco, G., del Rio, C. R., & Khazai, B. (2006). Post-disaster recovery dilemmas: Challenges in balancing short-term and long-term needs for vulnerability reduction. *Environmental Science and Policy, 9*(7–8), 607–613. https://doi.org/10.1016/j.envsci.2006.07.006

Ismaila, D., Majidb, T. A., Rooslic, R., & Samahd, N. A. (2014). Project management success for post-disaster reconstruction projects: International NGOs perspectives. *Procedia Economics and Finance, 18*(2014), 120–127. https://doi.org/10.1016/s2212-5671(14)00921-6

Johnson, C., & Lizarralde, G. (2012). Post-disaster housing and reconstruction. In *International encyclopedia of housing and home* (pp. 340–346). https://doi.org/10.1016/B978-0-08-047163-1.00046-1

Johnson, K., Depietri, Y., & Breil, M. (2016). Multi-hazard risk assessment of two Hong Kong districts. *International Journal of Disaster Risk Reduction, 19*(2016), 311–323. https://doi.org/10.1016/j.ijdrr.2016.08.023

Jordan, E. (2012). *Pathways to community* recovery*: A qualitative comparative analysis of post-disaster outcomes*. University of Colorado.

Li, Q., Umaier, K., & Koide, O. (2019). Research on post-Wenchuan earthquake recovery and reconstruction implementation: A case study of housing reconstruction of Dujiangyan City. *Progress in Disaster Science, 4*, 10041. https://doi.org/10.1016/j.pdisas.2019.100041

Li, X., Lam, N., Qiang, Y., Li, K., Yin, L., Liu, S., & Zheng, W. (2016). Measuring county resilience after the 2008 Wenchuan earthquake. *International Journal of Disaster Risk Science, 7*(4), 393–412. https://doi.org/10.1007/s13753-016-0109-2

Lizarralde, G. (2004). *Organisational system and performance of post-disaster reconstruction projects*. Universite de Montreal.

Lu, Y., & Xu, J. (2015). Comparative study on the key issues of postearthquake recovery and reconstruction planning: Lessons from the United States, Japan, Iran, and China'. *Natural Hazards Review, 16*(3), 04014033. https://doi.org/10.1061/(asce)nh.1527-6996.0000172

Lyons, M. (2009). Building back better: The large-scale impact of small-scale approaches to reconstruction. *World Development, 37*(2), 385–398. https://doi.org/10.1016/j.worlddev.2008.01.006

Marto, R., Papageorgiou, C., & Klyuev, V. (2018). Building resilience to natural disasters: An application to small developing states. *Journal of Development Economics, 135*(November 2018), 574–586. https://doi.org/10.1016/j.jdeveco.2018.08.008

Moatty, A. (2017). Post-flood recovery: An opportunity for disaster risk reduction? In F. Vinet (Ed.), *Floods* (Vol. 2, pp. 350–363). Elsevier. https://doi.org/10.1016/B978-1-78548-269-4.50023-8

Okada, T., Howitt, R., Haynes, K., McAneney, J., & Bird, D. (2018). Recovering local sociality: Learnings from post-disaster community-scale recoveries. *International Journal of Disaster Risk Reduction, 31*(2018), 1030–1042. https://doi.org/10.1016/j.ijdrr.2018.08.010

Okamoto, K., & Ishikawa, Y. (2013). Traditional wisdom and modern knowledge for the earth's future. In J. Noritaka Yagasaki (Ed.), *International geographical union Kyoto regional conference*. Springer.

Opdyke, A. (2017). Resilient and sustainable infrastructure systems: A comparative analysis of post-disaster shelter coordination, stakeholder participation, and training [University of Colorado Boulder]. In *ProQuest dissertations and theses*. https://doi.org/10.1108/ijdrbe-10-2017-0060

Patel, S., & Hastak, M. (2013). A framework to construct post-disaster housing. *International Journal of Disaster Resilience in the Built Environment, 4*(1), 95–114. https://doi.org/10.1108/17595901311299026

Platt, S. (2018). Factors affecting the speed and quality of post-disaster recovery and resilience. In S. Ó. Rajesh Rupakhety (Ed.), *Earthquake engineering and structural dynamics in memory of Ragnar Sigbjörnsson* (pp. 369–403). Springer. https://doi.org/10.1007/978-3-319-62099-2

Powell, P. J. (2011). Post-disaster reconstruction: A current analysis of Gujarat's response after the 2001 earthquake. *Environmental Hazards, 10*(3), 279–292. https://doi.org/10.1080/17477891.2011.597839

Rani, W. N. M. W. M., Nifa, F. A. A., Ismail, M. N., & Khalid, K. N. (2017). Planning for post disaster recovery: Lesson learnt from flood events in Kelantan Malaysia. In *AIP Conference Proceedings* (Vol. 1891, No. 1, p. 020143). https://doi.org/10.1063/1.5005476

Sadiqi, Z., Trigunarsyah, B., & Coffey, V. (2017). A framework for community participation in post-disaster housing reconstruction projects: A case of Afghanistan. *International Journal of Project Management, 35*(5), 900–912. https://doi.org/10.1016/j.ijproman.2016.11.008

Sadri, A. M., Ukkusuri, S. V., Lee, S., Clawson, R., Aldrich, D., Nelson, M. S., Seipel, J., & Kelly, D. (2018). The role of social capital, personal networks, and emergency responders in post-disaster recovery and resilience: A study of rural communities in Indiana. *Natural Hazards, 90*(3), 1377–1406. https://doi.org/10.1007/s11069-017-3103-0

Supernaw, S. A. (2019). *Impact of project management methodology use in post-disaster recovery and mitigation projects by North Carolina emergency managers* (Issue February). Northcentral University.

Tanwattana, P. (2018). Systematizing Community-Based Disaster Risk Management (CBDRM): Case of urban flood-prone community in Thailand upstream area. *International Journal of Disaster Risk Reduction, 28*(June 2018), 798–812. https://doi.org/10.1016/j.ijdrr.2018.02.010

The Centre for Research on the Epidemiology of Disasters (CRED). (2018). *Natural disasters*.

The Centre for Research on the Epidemiology of Disasters (CRED) and The UN Office for Disaster Risk Reduction (UNISDR). (2017). Economic losses, poverty & disasters 1998–2017. In *CRED UNISDR* (Vol. 01).

Tran, T. A. (2015). Post-disaster housing reconstruction as a significant opportunity to building disaster resilience: A case in Vietnam. In *Natural hazards* (Vol. 79, Issue 1, pp. 61–79). https://doi.org/10.1007/s11069-015-1826-3

Vahanvati, M., & Mulligan, M. (2017). A new model for effective post-disaster housing reconstruction: Lessons from Gujarat and Bihar in India. *International Journal of Project Management, 35*(5), 802–817. https://doi.org/10.1016/j.ijproman.2017.02.002

Wu, J. Y. (2003). A comparative study of housing reconstruction after two major earthquakes: The 1994 Northridge earthquake in the United States and the 1999 Chi-Chi earthquake in Taiwan. In *ProQuest dissertations and theses* (Vol. 28, Issue August). Texas A&M University.

Xu, P., Lu, X., Zuo, K., & Zhang, H. (2014). Post-Wenchuan earthquake reconstruction and development in China. In K. T. L. Naim Kapucu (Ed.), *Disaster and development examining global issues and cases*. Springer. https://doi.org/10.1007/978-3-319-04468-2_25

Yang, X., & Zhang, Q. (2013). Seismic spatial information grid: Applications of geo-informatics in earthquake disaster management. In *International Conference on Geo-Informatics in Resource Management and Sustainable Ecosystem, 399 PART I* (pp. 397–406). https://doi.org/10.1007/978-3-642-41908-9_41

Chapter 2
Research Applied Methodology

Abstract This section outlines the specific methods used within this research. The overall purpose of the chapter is to provide a robust rationale for selecting the research process to meet the research aim and objectives. The research design of this book is proposed based on the framework proposed by *Saunders et al. (2007)* as the topic of this chapter (Saunders et al., 2007). Therefore, the methods applied in data collection and analysis are presented after describing the research design process. Followingly the chapter ends by explaining how the research's validity and reliability have been guaranteed.

Keywords Research method · Data collection · Data analysis

2.1 Research Strategy and Procedure

As this study attempts to clarify the links between variables, it is categorized as explanatory research (Flores Salas, 2016). This study decomposed the research process into seven steps as follows:

(1) Research problem identification and proposing the research questions through the literature review (first-round).
(2) Categorization of the relevant literature.
(3) Extraction of the influential factors in disaster recovery projects after socio-natural disasters through the literature review (second-round).
(4) Verification of the extracted influential factors.
(5) Exploring Interaction among the influential factors and success dimensions.
(6) The implication of the model in reality.
(7) Presentation of the results and conclusions.

As this study proposed six questions, Fig. 2.1 graphically illustrates the links between the research's suggested methods and the study questions. Additionally, the authors have considered the time-based expected results of the study, which can be measured in each step of the research process. Based on the research conducted by

H. Bahmani et al., *Socio-Natural Disaster Recovery Projects*, Resilient and Friendly Living Environments, https://doi.org/10.1007/978-981-95-0113-7_2

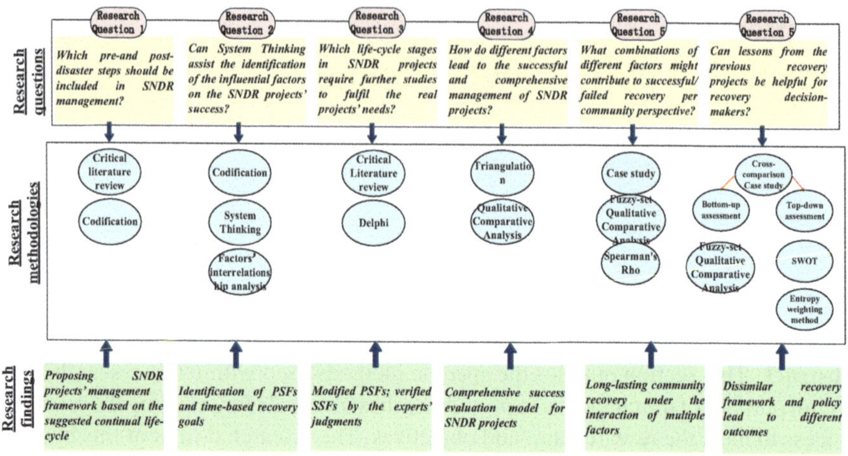

Fig. 2.1 The links between the research's questions and research methodologies

Yilmaz et al., (2013), this framework divides the primary purpose of this study into several sub-objectives. Figure 2.2 shows the research process schematically.

2.2 Data Collection Methods

2.2.1 *Critical Literature Review*

The research commences with searching for reliable and relevant academic resources. The resources include theoretical and empirical findings related to socio-natural disaster recovery projects. The resources have been chosen from thousands of electronic academic resources available at the time of this research. The authors found the resources through academic websites such as Science Direct, ProQuest, and Springer by initially searching these websites using the keywords "disaster recovery" AND "socio-natural disasters". The published resources during 2003–2020 have been recognized and listed in Table 2.1.

However, the authors decided to apply for critical review due to the high number of short-listed resources resulting from the research's goal to comprehensively cover the recovery projects after socio-natural disasters. This method, which extensively and critically evaluates the resources, does not simply focus on the narration of the literature. The result of this methodology might be the starting point for the rest of the research as it produces a hypothesis or a conceptual framework (Grant & Booth, 2009). As a result, "post-natural disaster recovery" and "post-disaster reconstruction" were the additional keywords used to refine the English-written resources. The most relevant journals from several publishers have been identified. The resources from books, dissertations, conference papers, and academic reports have also been chosen.

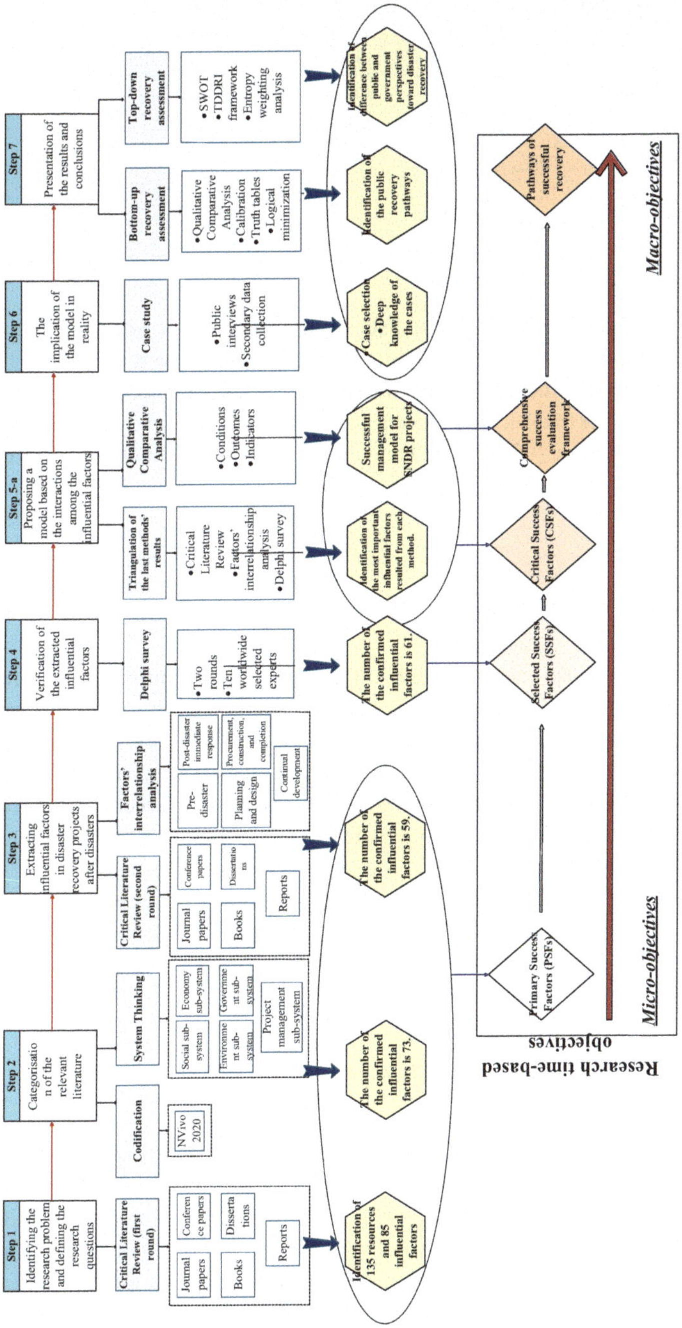

Fig. 2.2 Research methodology roadmap

Table 2.1 Search engines' results- Refined results based on the primary keywords search

Academic search engine	Number of chapters	Number of articles	Number of conference papers	Number of books	Total number
Science Direct	3844	13,185	198	44	17,271
Springer	17,546	9093	1925	287	28,851
ProQuest	*Number of dissertations*				
	3126				
Total number					49,248

The authors tried to select resources that discuss different types of natural hazards and simultaneously include global cases. While Table 2.2 walks the reader through the types of this study's resources, countries of case studies, and disaster types. Figure 2.3 shows the frequency of the selected journal papers. Finally, the study comprised 135 resources published from 2003 to 2020 and covered different countries and various disaster types.

2.2.2 *Delphi Survey*

The authors decided to use experts' judgment to check the reliability of the extracted success factors from the literature review. Therefore, the Delphi method is one of the best and most well-known operational techniques widely used (Pomponi et al., 2019). Delphi can be applied to extract information that can build consensus upon the primary extracted factors (Hasson et al., 2000). The identified success factors can be validated and refined by taking advantage of unbiased judgments from anonymous panellists. Additionally, sharing the last round's results with the panellists provides the research with an excellent opportunity to reach a reliable consensus (Diamond et al., 2014). Moreover, due to the panellists' geographical dispersion, the Delphi survey is a practical solution to save time (de Brito et al., 2017; Opdyke et al., 2020).

2.2.3 *Case Study; Structured Interviews and Official Report*

The case study strategy is mostly a suggested methodology if the researchers tend to gain a rich understanding of the research context. This technique is able to answer the research questions such as "why?", "what?" and "how?" (Saunders et al., 2007). The case studies are applied to generalize the findings to more extensive and similar contexts; however, the researcher must carefully select the cases and the data analysis methodology to prevent further defeats in generalization (Yin, 2003). Data can

Table 2.2 The categorization of the study's resources

Resource		Num. of publications	Case study	Disaster type
Books		7	China, Japan, USA, Iran, Pakistan, Thailand, Italy, Chile, New Zealand, Turkey, Sri Lanka	Earthquake (5), Tsunami (1)
Dissertations		19	Japan, USA, Taiwan, China, India, Japan, Indonesia, Sri Lanka, Turkey, Iran, El Salvador, Haiti, USA	Earthquake (15), Flood (3), Tsunami (3)
Others		1	USA	-
Conference papers		9	Multidimensional case comparison, Malaysia, Turkey, China	Earthquake (3), Flood (1), Tsunami (2)
Journal papers		99		
Journal papers' publishers	*ASCE Library (A)*	5	India, Turkey, USA, Iran, China, Japan	Earthquake (3), Hurricane (1)
	De Gruyter (B)	1	USA	-
	Elsevier (C)	60	Philippines, Peru, Pakistan, Indonesia, Sri Lanka, Afghanistan, Haiti, Vietnam, Peru, Burkina Faso, China, Indonesia, El Salvador, Palestine, Honduras, American continent, Pakistan, Bangladesh, Iran, Philippines, Haiti, Burkina Faso, Turkey, Nepal, New Zealand, Australia, Japan, USA, Thailand, India, Vanuatu, Tuvalu, Kiribati, Italy, Nepal	Earthquake (28), Flood (10), Tsunami (8), Typhoon (3), Hurricane (1), Storm (1), War (4), Cyclone (3), Debris flow (1), Volcanic eruption (1)
	Emerald Publishing (D)	16	India, China, Cook Islands, Sri Lanka, Malaysia, Turkey, India	Earthquake (3), Flood (2), Tsunami (4)
	MDPI (E)	2	Philippines	Typhoon
	Springer (F)	6	USA, Vietnam, China, Sri Lanka	Earthquake (3), Flood (1), Tsunami (2), Storm (2), Cyclone (1)
	Taylor and Francis Group (G)	6	India, China, USA	Earthquake (1), Tsunami (1), Hurricane (1)
	University Press (H)	3	Iran, India, USA, Japan	Earthquake (2), Tsunami (2), Hurricane (1)

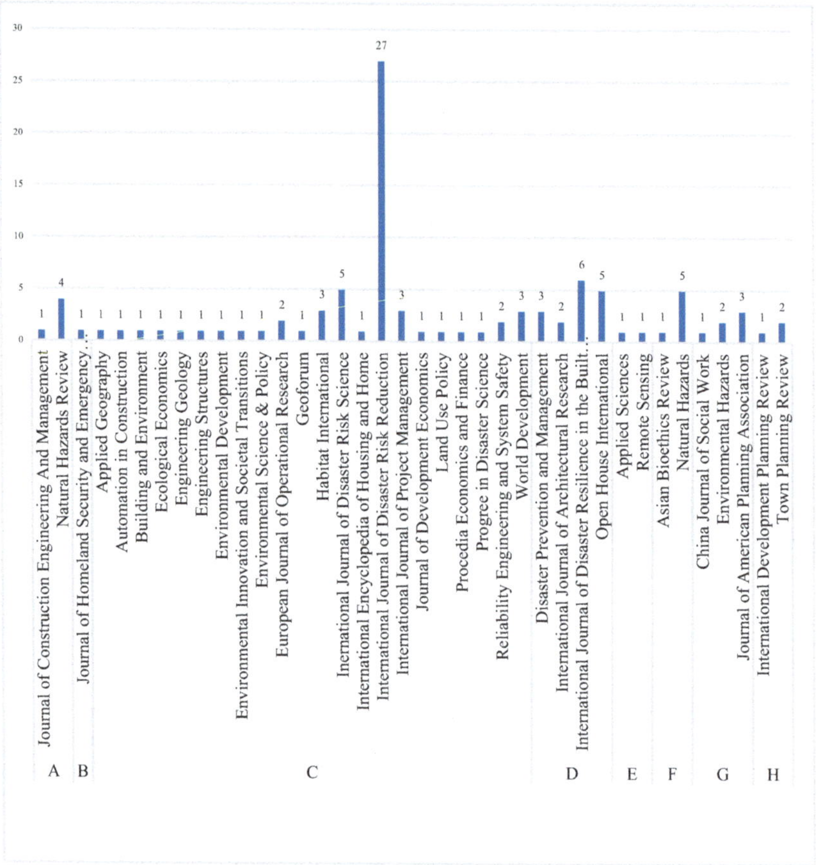

Fig. 2.3 Journal papers' frequency

be collected through various techniques such as interviews, questionnaires, observations, and secondary documents. However, some studies have suggested the application of triangulation methodology to verify the case study's findings. Applying this technique, the researcher must collect data from one study using different methods (Saunders et al., 2007). This study has chosen two cases from the severely affected countries, Iran and China, to conduct a cross-comparison analysis. The case studies' data have been collected through structured public interviews in the selected neighbourhoods in Bam, Iran. Two trained interviewers were conducted who were also asked to do observations. The interviewees were selected based on the research criteria and the questions to evaluate the public living conditions before and after the earthquake, Bam, 2003.

In the case of the Wenchuan earthquake, official reports by the Sichuan government from 2007 to 2012 were also accessed through the yearbooks of the province.

This data helped us to construct the top-down disaster recovery assessment and show the recovery trends and the government's focus over the recovery years.

2.3 Data Analysis Methods

2.3.1 Codification

After selecting the resources, Mendeley has been utilized as the research's documentation system. Per the multidimensional nature of SNDR projects covering various topics, throughout the application of NVivo 2020, a codification of the research library has been done. These codes are roadmaps to create two sets of categorizations based on "projects management stages" and "project management life cycle" and attempt to reasonably consider the activities and their different required duration.

2.3.2 System Thinking and Factors' Interrelationship Analysis

The traditional thinking method breaks a project into parts that neglect interrelationships, while system thinking attempts to track the pattern of changes. System theory includes the principles, models, and laws necessary to understand the complex interrelationships and interdependencies between the components of a complex system (Fan et al., 2015). SNDR projects can be seen as systems that consist of numerous components. These projects are influenced by a complex environment, causing different difficulties and making the recovery project a tough job to control (Lizarralde, 2004). System thinking, "*seeing the forest for the trees*", was applied in this research to ease understanding of the intricate nature of SNDR projects. Throughout the application of system thinking, five sub-systems, 16 groups, and 73 factors influencing SNDR projects' success have been coded by NVivo 2020. The frequency of the categorized factors based on the literature review was then analyzed. Two rounds of factor interrelationship analysis were conducted to check the accuracy and redundancy of the extracted influential factors. This process can also ease the data collection and calculation of future studies. The influential factors were assessed throughout the continuous life cycle of SNDR projects.

2.3.3 Triangulation

The authors applied triangulation methodology to gain refined perspectives on the influential factors that influence the success of SNDR projects. This methodology benefits from using multiple research methods simultaneously to understand the construct (Mostafa & Sallam, 2005). Reliability checks of different resources and accuracy tests of the researcher's interpretation are some of the advantages of this method resulting from the simultaneous use of qualitative and quantitative methods (Heath, 2015). However, the application of triangulation can be costly and time-consuming. In this research, the authors have chosen the Delphi method to analyze the research quantitively, while a critical literature review alongside the factors' inter-relationship analysis qualitatively evaluates the data. Later, triangulation categorized the results of the last-mentioned stages and set acceptance criteria for each method to select the top-ranked factors.

2.3.4 Narrative Comparative Analysis (NCA) and Strengths, Weakness, Opportunities, Threats Analysis (SWOT) Analysis

The study benefited from applying NCA and compared the earthquake recovery projects within two countries, Iran and China. The cases have been narratively explained and compared, considering the proposed SNDR success model and continuous life cycle. The research applies SWOT analysis, a strategic planning method that provides evaluation tools for each country to assess internal and external conditions that negatively and positively have affected the recovery process in each case.

2.3.5 Qualitative Comparative Analysis (QCA) and Spearman's Rho Test

The results obtained from the triangulation method have been interpreted using the QCA method. This method attempts to compare cases, which was first introduced by *Charles Ragin* in 1987 (Befani, 2016). QCA, which uses the advantages of traditional qualitative and quantitative research methods within-case comparison, has been nominated as the third way to do social research. However, the logic behind this method differs from the traditional qualitative and quantitative techniques as it benefits from set theory and Boolean algebra (Devers et al., 2013). Generally, QCA defines which factors/combinations of factors might affect the results within the content of carefully selected cases (Meur et al., 2006). Applying QCA in SNDR projects may enable the researchers to identify where, when, and under what circumstances the project can be managed successfully. Additionally, the Spearman correlation coefficient test running in SPSS 26 has assessed correlations between the success factors.

2.3.6 *Entropy Weighting Method*

The entropy method is employed to quantify the degree of disorder or uncertainty across various assessment domains to evaluate disaster recovery complexity. This study determined the importance of several recovery indicators across years and cases using multidimensional entropy analysis. The entropy technique takes a look at how the data is distributed and uses that knowledge to give each indicator an objective weight. This way, the recovery evaluation is skewed toward dimensions with more variability. Reducing bias in indicator selection and improving the reliability of comparative analyses, this strategy offers a data-driven way to evaluate recovery performance. The results of this analysis were later presented in regional maps to show recovery trends over the Wenchuan reconstruction years using QGIS 3.38.0.

2.4 Validity and Reliability of the Methodology

In order to provide credibility to the results, it is vital to validate the research methods. The validation of this research has been assessed as follows (Befani, 2016).

(1) **Construct validity**: The researchers have collected the data from different data collection methods to obtain construct validity. The relevant literature has been selected based on the criteria explained previously. Additionally, the selected resources have been read twice. The Delphi survey has been conducted and analyzed based on the research guidelines. Moreover, the panel members have been carefully selected to meet the study's requirements. Moreover, to guarantee the research's ability to generalize the findings, the case studies have been selected based on the predefined criteria.
(2) **Internal validity**: The accuracy of collected data from the case studies and analyzed by QCA has been calibrated based on the precise rules. This process proved that the research avoids bias against particular criteria or data.
(3) **External validity**: The study collected data from multiple cases with different backgrounds. However, based on the logic of QCA, it is necessary to choose the cases with some similarities meanwhile considering their different characteristics (Meur et al., 2006).
(4) **Reliability**: The study has been conducted by establishing clear protocols covering all the research steps. These protocols include literature selection criteria, codification guide based on system thinking and factors' interrelationship analysis, guidelines for conducting and analyzing the Delphi survey, case study selection criteria, interview guidelines, participant selection, and data calibration rubrics of the case studies.

2.5 Chapter Summary

The chapter provides detailed information about the research methodology that will guide the data collection, processing, and analysis process. Surveys with experts and public and mixed-method (qualitative and quantitative data) were applied in this study. A critical literature review, Delphi, case study, and structured interviews were the instruments used for data collection. In data analysis, alongside system thinking, factors' interrelationship analysis based on the SNDR projects' life cycle, codification, triangulation, QCA, and Spearman's Rho test have been used to analyze data. The chapter was finalized by explaining the validity and reliability of the selected methods.

References

Befani, B. (2016). *Pathways to change: Evaluating development interventions with Qualitative Comparative Analysis (QCA).* Elanders Sverige AB.

de Brito, M. M., Evers, M., & Höllermann, B. (2017). Prioritization of flood vulnerability, coping capacity and exposure indicators through the Delphi technique: A case study in Taquari-Antas basin, Brazil. *International Journal of Disaster Risk Reduction, 24*(June), 119–128. https://doi.org/10.1016/j.ijdrr.2017.05.027

De Meur, G., Bursens, P., & Gottcheiner, A. (2006). MSDO/MDSO revisited for public policy analysis. In B. Rihoux & H. Grimm (Eds.), *Innovative comparative methods for policy analysis: Beyond the quantitative-qualitative divide* (Issue January, pp. 1–344). https://doi.org/10.1007/0-387-28829-5

Devers, K. J., Lallemand, N. C., Burton, R., Kahwati, L., Mccall, N., & Zuckerman, S. (2013). *Using Qualitative Comparative Analysis (QCA) to study patient-centered medical homes* (Issue September).

Diamond, I. R., Grant, R. C., Feldman, B. M., Pencharz, P. B., Ling, S. C., Moore, A. M., & Wales, P. W. (2014). Defining consensus: A systematic review recommends methodologic criteria for reporting of Delphi studies. *Journal of Clinical Epidemiology, 67*(4), 401–409. https://doi.org/10.1016/j.jclinepi.2013.12.002

Fan, Y., Li, Z., Pei, J., Li, H., & Sun, J. (2015). Applying systems thinking approach to accident analysis in China: Case study of "7.23" Yong-Tai-Wen High-Speed train accident. *Safety Science, 76*(2015), 190–201. https://doi.org/10.1016/j.ssci.2015.02.017

Flores Salas, A. (2016). *Evaluation of the use of lightweight concrete panels for post disaster house reconstruction using building information modelling.* University of Manchester.

Grant, M. J., & Booth, A. (2009). A typology of reviews: An analysis of 14 review types and associated methodologies. *Health Information and Libraries Journal, 26*(2), 91–108. https://doi.org/10.1111/j.1471-1842.2009.00848.x

Hasson, F., Keeney, S., & McKenna, H. (2000). Research guidelines for the Delphi survey technique. *Journal of Advanced Nursing, 32*(4), 1008–1015.

Heath, L. (2015). Triangulation: Methodology. In *International encyclopedia of the social & behavioral sciences: Second edition* (2nd ed., Vol. 24, pp. 639–644). Elsevier. https://doi.org/10.1016/B978-0-08-097086-8.44059-6

Lizarralde, G. (2004). *Organisational system and performance of post-disaster reconstruction projects.* Universite de Montreal.

Mostafa, M., & Sallam, I. (2005). Triangulation methodology: Can it be used to improve environmental assessment strategies? In *4th Australian Life Cycle Assessment Conference, January 2011.*

Opdyke, A., Goldwyn, B., & Javernick-Will, A. (2020). Defining a humanitarian shelter and settlements research agenda. *International Journal of Disaster Risk Reduction, 52*(March 2020), 101950. https://doi.org/10.1016/j.ijdrr.2020.101950

Pomponi, F., Moghayedi, A., Alshawawreh, L., D'Amico, B., & Windapo, A. (2019). Sustainability of post-disaster and post-conflict sheltering in Africa: What matters? *Sustainable Production and Consumption, 20,* 140–150. https://doi.org/10.1016/j.spc.2019.06.007

Saunders, M., Lewis, P., & Thornhill, A. (2007). Research Methods for Business Students. In *Pearson Education Limited 2.*

Yilmaz, D. G., Meding, J. V., & Erk, G. K. (2013). A theoretical approach to the design of a survey instrument in post-disaster reconstruction: Defining indicators for a human-based study in rural built-environment. *International Journal of Architectural Research, 7*(3), 40–56.

Yin, R. K. (2003). *Case study research design and methods* (3rd ed.). Applied Social Research Methods Series.

Chapter 3
Application of System Thinking in Socio-Natural Disaster Recovery Projects

Abstract System thinking, which can reduce the complexity of systems and clarify the interrelationships of the components, has been applied to provide an organized understanding of socio-natural disaster recovery projects. System thinking divided SNDR projects into five sub-systems through a qualitative analysis of the research's selected resources coded by NVivo 2020. As the project management stage was recognized as the most crowded sub-system, an integrated management framework has been put forward throughout the simultaneous application of the continuous lifespan and project stages. The most significant observation of this study shows the significant role of system thinking in organizing SNDR projects. Moreover, the necessity of applying a continuous lifespan to address projects' long-term to manage future socio-natural disasters has been highlighted.

Keywords System thinking · Disaster recovery life-cycle · Pre-disaster stage · Post-disaster stage

3.1 System Thinking's Basic Principles and Its Application in SNDR Projects

Considering systems as nonlinear and dynamic feedback loops that have a group of elements with relations between them and their environment, systems thinking is a method to see a whole (Rani et al., 2017). In research conducted by Arnold and Wade (2015), the definition of system thinking has been assessed through several studies. The majority of that study's resources defined system thinking as a method to see the whole (rather than parts) that can be used in dynamic and complex behaviour to simplify and define the interrelationships/ interconnections among the systems' items. However, those scholars believed that system thinking is "*a set of synergistic analytic skills used to improve the capability of identifying and understanding systems, predicting their behaviors, and devising modifications to them in order to produce desired effects. These skills work together as a system*" (Arnold & Wade, 2015). Another study differentiated "*thinking about a system*" from "*system thinking*"

H. Bahmani et al., *Socio-Natural Disaster Recovery Projects*, Resilient and Friendly Living Environments, https://doi.org/10.1007/978-981-95-0113-7_3

and claimed that the former is an ad hoc way as each person thinks differently about the system. Subsequently, that study explained that system thinking is in accordance with the conceptual patterns of organizations (Cabrera et al., 2008).

In accordance with the context analysis, the use of system thinking in project management is prevalent. System thinking has been applied in research analyzing the main causes of railway accidents in China (Fan et al., 2015). Similarly, the causes of construction accidents have been analyzed through system thinking, and those factors have been categorized into several sub-systems and groups (Zhang et al., 2020). Another study has applied system thinking to identify the cause and effect of flood management in Pakistan (Rehman et al., 2019).

The last examples illustrate that SNDR projects can be seen as systems that consist of numerous components. Since SNDR projects are intricate and may continue for a long time, appropriate management requires significant knowledge of the content of activities that need to be planned and implemented. These projects are influenced by a complex environment, causing different difficulties and making the recovery project a tough job to control. Hence, system thinking, which is able to simplify the complexity within systems, can efficiently identify and group recovery tasks. Seeing SNDR projects as a system, considering branches for the system is essential to utilise the system thinking concept fully.

According to the literature review, the researchers proposed the following framework by applying system thinking. Five sub-systems, the main branches of this framework, include the multidimensional aspects influencing SNDR projects. Social, environmental, and economic sub-systems are derived from the sustainability concept and resemble this fact that not only should the recovery projects after socio-natural disasters focus on rebuilding and improving buildings, but also recovery of the business and community structures should be considered. However, to provide a comprehensive perspective consisting of the most concerning topics in SNDR projects, the study added government and project management sub-systems to the systematic framework. Additionally, the last two sub-systems are one of the most problematic sections since numerous activities can be placed there. Subsequently, the groups resulting from the division of the sub-systems have been added. The groups work as indicators to determine the influencing factors and place each element appropriately. The system thinking framework is demonstrated graphically in Fig. 3.1.

Throughout the critical content analysis of the research selected resources, 85 influential factors were extracted, shown in Fig. 3.2. The initial factors have been placed in five sub-systems based on the literature review's mentioned topics. This categorization endeavoured to address most of the common concerns in the SNDR projects.

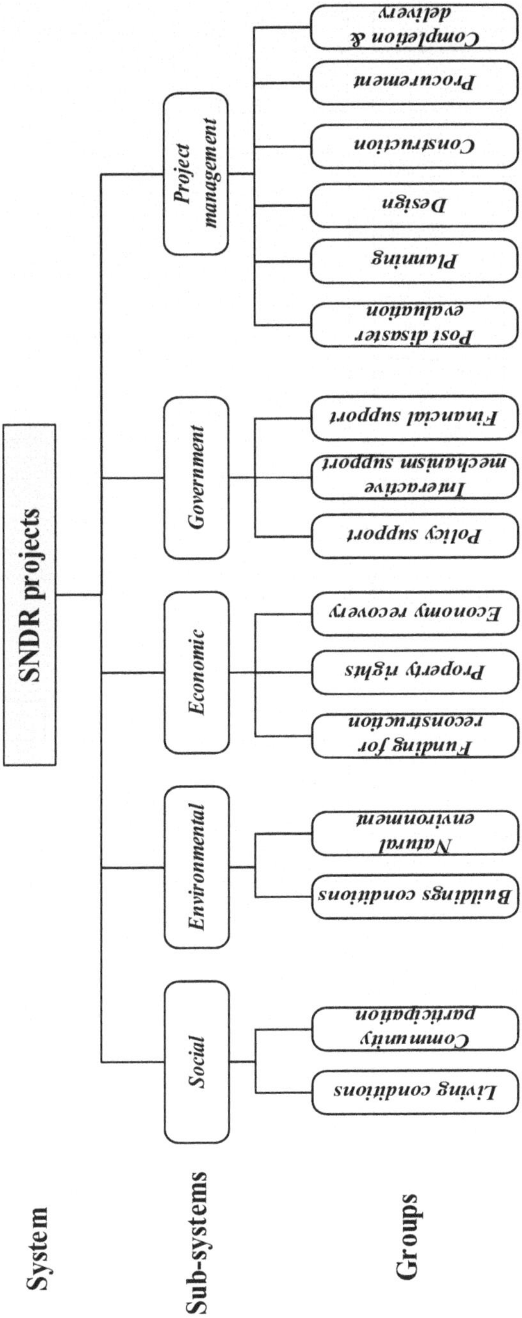

Fig. 3.1 SNDR project's components- application of system thinking

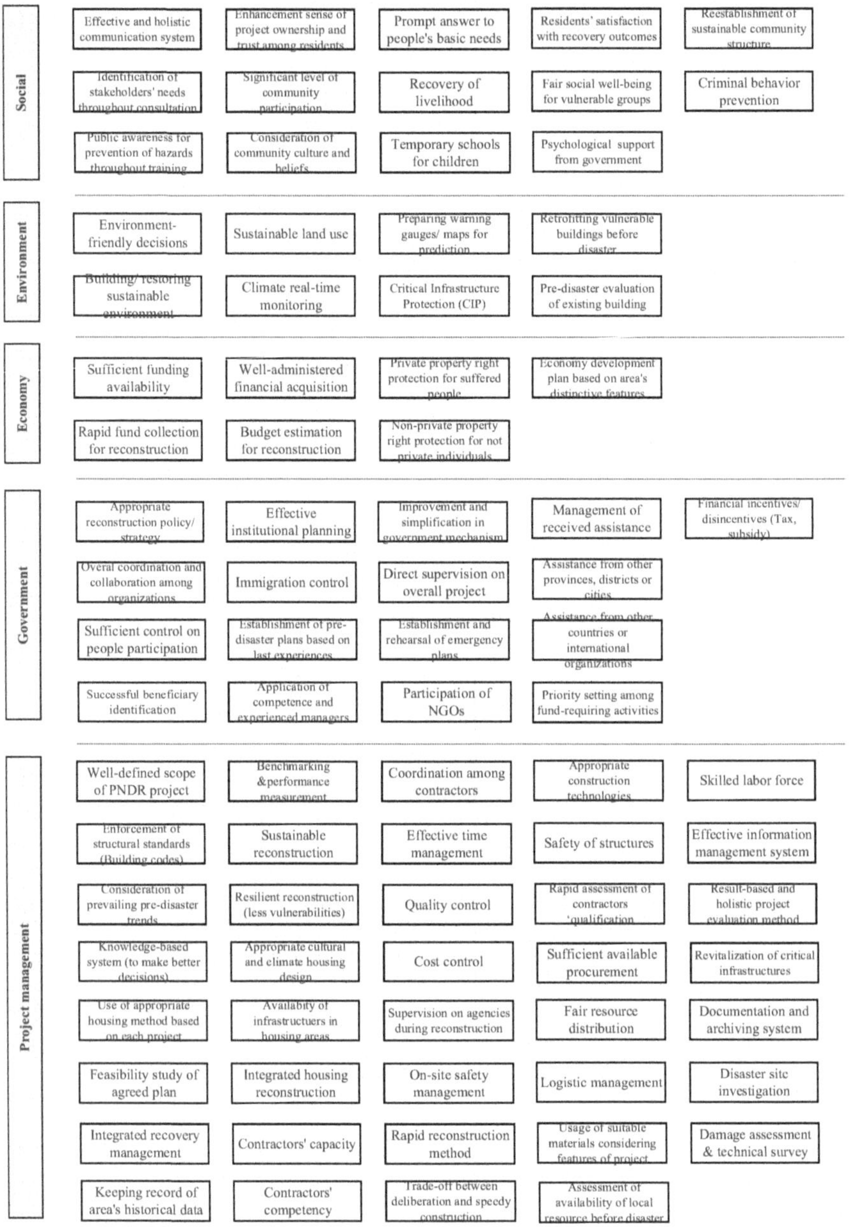

Fig. 3.2 Initial identified factors from the context analysis

3.2 Decomposition of SNDR Projects' Lifespan and Stages

Recovery projects after a socio-natural disaster should be seen as a portfolio consisting of heterogeneous but dependent projects varying from physical recovery to cultural and economic development (Lizarralde, 2004). Furthermore, detailed and clear management planning is a requisite for recovery success. Following the traditional definition of projects, especially construction projects, linear life cycle management has been mostly used in recovery projects. This short-term and linear life cycle that neglects long-term development is a demanding challenge that needs to be solved. A large body of evidence alongside Fig. 3.3 reveals that lessons learned, collected throughout the recovery project, will not be used as gained experience to prepare communities for the subsequent socio-natural disasters.

Several proposed life cycles for the management of SNDR projects acknowledged the increasing need for applying a practical management framework that can properly answer the different requirements of SNDR projects (Bosher et al., 2021; Vahanvati & Mulligan, 2017). However, for a more concise understanding of this concept, the studies must identify the sequences and importance of the life cycle phases. Moreover, a need exists to recognize and locate the tasks within the dynamic life cycle of recovery projects.

Additionally, the importance of simultaneous consideration of the pre-and post-disaster planning alongside the concept of comprehensive recovery without marginalizing intangible aspects is somewhat forgotten in some cases (Walker et al., 2017; He et al., 2021; Rodrigo & Wilkinson, 2021). On the other hand, segregated perspectives against pre- and post-disaster management (Ghaffarian et al., 2020, 2021; Sheykhmousa et al., 2019) may find a proper answer by proposing a systematic and continual lifespan. Generally, a growing body of literature discusses socio-natural disaster recovery projects' life cycle by focusing on "pre-disaster planning" and "post-disaster planning" (Kermanshachi et al., 2019; Xu & Lu, 2018).

3.2.1 Pre-Disaster Stage

More recent work shows that recovery needs previous plans to organize post-disaster activities (Xu et al., 2014). Se studies suggest that disaster management should shift from recovery-oriented to mitigation-oriented policies (Platt, 2018). Focusing on the advantages of establishing plans before a disaster, a pre-disaster investment of one dollar benefits about four dollars a post-disaster stage in case of an earthquake (He & Zhuang, 2016). However, effective pre-disaster planning relies heavily on the cooperation of people and experts from different fields (Edgington, 2011). Pre-disaster planning, called "*up-front planning*" (Johnson, 2006) or "*pre-planning phase*" (Vahanvati & Mulligan, 2017), has been focused on by many scholars due to its worthwhile effects on time and cost management and clear task identification,

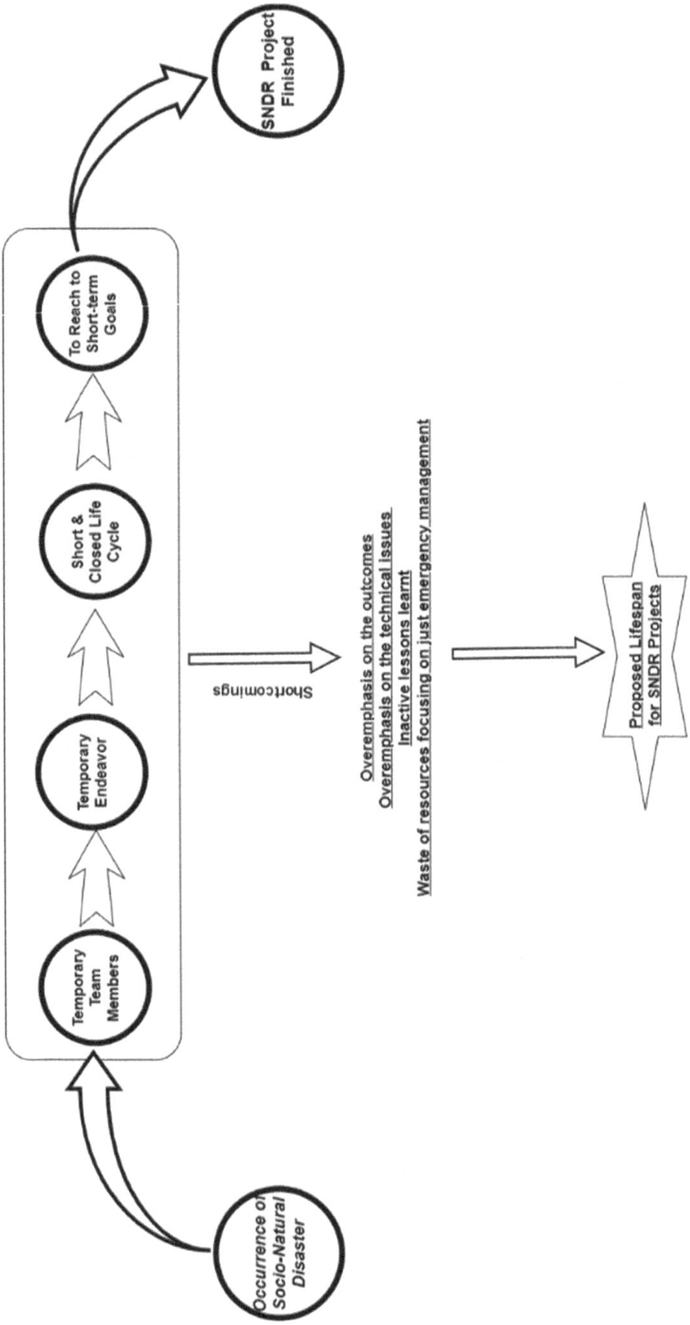

Fig. 3.3 Widely applied linear life cycle for disaster recovery management projects, adopted based on (Lizarralde, 2004)

which provide an opportunity to take more advantage of the window of opportunity after the disaster (Fan et al., 2018).

Overall, the results presented by some researchers largely ignore overlaps in explanations of the pre-disaster stage and divide it into three phases: preparedness, prevention, and prediction. Recent works have considered this issue and "mitigation" and "preparedness" as a pre-disaster stage in those studies (Altay & Green, 2006; Xu et al., 2014). In the lens of sustainable recovery, preparation has been regarded as disaster mitigation, and risk reduction strategy is a vital prerequisite for maintaining sustainable development (Ahmed & Charlesworth, 2015; Yilmaz et al., 2013). A number of categorizations have been proposed in the recent literature on mitigation measurements consisting of three categories: community protection works (dams, levees, and drainage systems), land-use practices, and building construction practices (Jordan, 2012). Based on the previous research, Fig. 3.4 renders the most mentioned subjects that should be thought about in pre-disaster planning.

3.2.2 Post-Disaster Stage

Evidence supports the necessity of updating the pre-established post-disaster recovery plans after the disaster occurrence based on first-hand information (Griffey, 2010; Kermanshachi et al., 2019). The post-disaster recovery process differs from project to project due to numerous factors such as the severity of the disaster, physical damages, casualty numbers, existing trends before the disaster, recovery assistance programs, access to resources, geographic features, population, economic growth, and policies (Mukherji, 2008; Xu et al., 2014).

In most of the read papers, post-disaster recovery consists of two parts; "Emergency response" and "recovery phase" (Chen & Booth, 2011; Han, 2014). On the other hand, recent research has emphasized the well-being of social, mental, and economic conditions as well as physical recovery (Han, 2014; Xiao et al., 2020). Within the same lens, a comprehensive post-disaster recovery consists of "soft recovery" and "hard recovery" (Akiyama, 2019). Soft recovery, also called non-structural recovery, comprises standards and regulations established to tackle socio-natural disasters. These regulations can be seen as land-use policies, risk analysis, insurance, and financial compensations. Hard recovery (structural recovery) includes architectural design and the construction of physical barriers against future hazards (Lu & Xu, 2015).

Although some previous research expresses recovery as a bounce back to normalcy, prevalent before the disaster, some researchers believe that the term "recovery" is mainly oversimplified by defining it as turning to normality (Jordan, 2012; Platt, 2018). Comparing these definitions of recovery, this study proposed two recovery phases; "Short-term recovery" and "long-term recovery". While the former deals with bouncing back to the pre-existing situation before the disaster, the latter focuses on developing a strategy to improve societies' resilience, equity, and sustainability (Mukherji, 2008). On the other hand, the recovery phase starts immediately

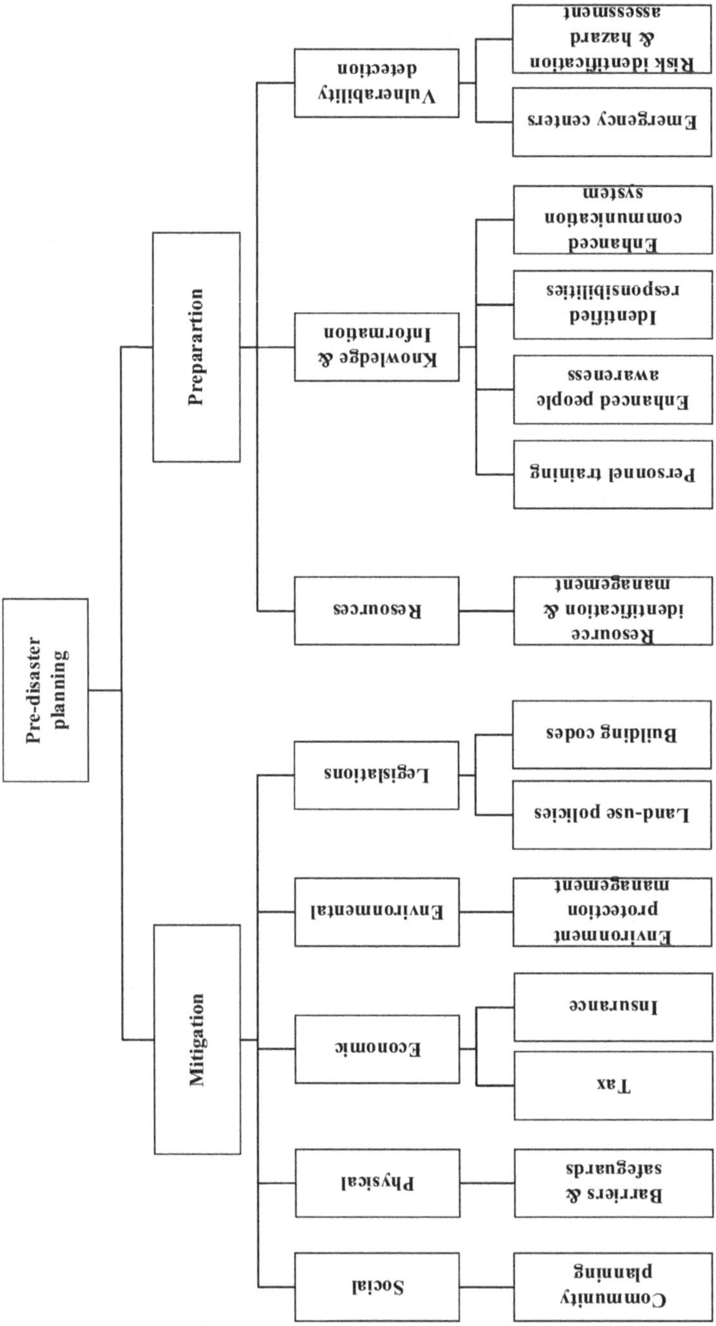

Fig. 3.4 Stages in pre-disaster planning

after a disaster, shows considerable overlaps with the emergency phase, and should be linked to development efforts (Liu et al., 2014; Platt, 2018). In this study, the emergency phase consists of all the activities starting just after a disaster, such as rescue, sheltering, restoring critical structures, and considering secondary hazards. These short-term activities can last for months. "Short-term recovery" has been counted as part of the "emergency" phase, while holistic development of a system has been categorized within the "recovery" stage. Additionally, there is irrefutable evidence suggesting that the term "recovery" is preferred over "reconstruction" because of the recent concerns about recovering multidimensionally (Akiyama, 2019). Figure 3.5 is this research's proposed model for the post-disaster recovery phase and its steps.

(1) **Hard (structural) recovery**

In the aftermath of a disaster, the reestablishment of structures is firmly connected to enhancing the general well-being of the residents. Reconstruction after a socio-natural disaster has two sides: reconstruction of the destroyed facilities and development. The latter topic has been discussed as a fundamental key for prevention and preparedness against future socio-natural disasters (Hidayat & Egbu, 2010). The recovery phase has also been divided into: "housing" and "infrastructures". Sheltering and temporary housing should be categorized in the emergency phase, while permanent housing should be planned in the recovery stage. Construction of memorial buildings (Ge et al., 2010) and protection against natural hazards (Du et al., 2016) are also part of structural

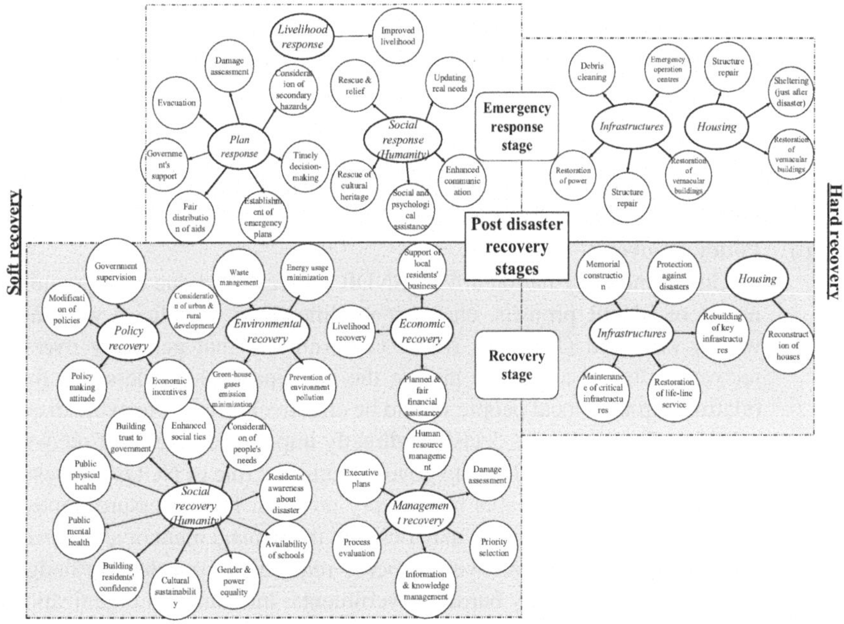

Fig. 3.5 Planned activities and phases of the post-disaster recovery stage

recovery; however, residents' overconfidence in the protective structures should be appropriately managed (Jordan et al., 2016).

(2) **Soft (non-structural) recovery**

(a) **Management recovery**

In the aftermath of a disaster, multitudinous topics will arise that necessitate the existence of a management body. Alongside strategic decision-making, namely, rules and legislation, management is essential to ensure the validity of the established rules. Management of the population and preventing the population's migration from hit areas to other places and vice versa should also be considered (Ge et al., 2010; Zhang & Peacock, 2010). Additionally, experience in coping with post-disaster recovery challenges, also known as lessons learnt from the past, has been emphasized by several scholars (Liu & Han, 2009; Meding et al., 2014).

(b) **Social recovery (humanity)**

The evidence from the last two decades strongly supports the severe effects of socio-natural disasters both people's physical and mental situations (Ge et al., 2010; Liu & Han, 2009). A study nominated prerequisites for successful social recovery in SNDR projects as stable institutions, the supply of human resources, and financial support. Moreover, a family's income, health conditions, availability of education, personal fulfilment, and social conditions have been recognized as life satisfaction factors (Ding et al., 2019).

(c) **Economic recovery**

In general, the recovery of livelihood (local business) after a disaster is the researchers' main concern when it comes to economic reestablishment (Edgington, 2017; Guo, 2012). Within the same context, "*sustainable livelihood* analyzing" was proposed to boost a community's capabilities to cope with disasters (Han, 2014). Restoring the economy was considered necessary to create a thriving temporary town and fulfil the reconstruction's objectives (Dikmen, 2007; Okamoto & Ishikawa, 2013).

(d) **Policy recovery**

Government regulation affects SNDR projects, and due to the requirements of SNDR projects, changes/ modifications in policies were also widely observed (Johnson, 2006). Governments that generally oversee recovery after disasters are among the principal stakeholders and their relationship with local people should be effectively framed to avoid dissatisfaction. Governments' decisions directly impact the success of recovery plans (He, 2019). In addition to governments, the role of NGOs (domestic/ international) is undeniable in recovery projects. High pressure imposed by disasters on governments may obstruct the management of all recovery stages. Moreover, the recovery process requires a considerable budget, which would financially burden governments. International organizations such as the World Bank, United States Office for Disaster Risk Reduction, and International Federation of Red Cross and Red Crescent Societies can efficiently smooth the recovery pathways by financing projects and providing technical assistance, which also can boost the hosting countries' mitigation and preparedness capabilities. However, inflexible project evaluation measurements can be troublesome in the context of long-term and intangible recovery (Ganapati & Mukherji, 2014).

(e) **Environmental recovery**

Most of the current evidence reveals environmental damage. SNDR projects should focus on ecological-friendly decisions and, at the same time, attempt to restore the damaged environment. Environmental management (mitigation activities) has been suggested to continue after short-term recovery to become part of the development plan (Wu, 2003). Afforestation reduces the probability of landslides and multi-purpose reservoirs. Additionally, constructing ecological infrastructure can effectively restore the damaged environment (Ge et al., 2010).

3.2.3 Integration Stage for Continual Tasks

To locate the mentioned activities in the last sections in their appropriate place within SNDR projects' lifespan, some activities can be grouped into pre-disaster and post-disaster stages. These activities, which should be continuously followed, have set an elastic phase named "Integration stage". The first item in this category, disaster vulnerability reduction, focuses on the holistic perspective considering all aspects of society, from the building structure, economy, livelihood, social well-fare, and environment, to improve the SNDR process comprehensively. Within this lens, the SNDR project has often been seen as an opportunity to Build Back Better, boost productivity, and attract more investment (Edgington, 2017; Rose, 2011; Siriwardhana et al., 2012). Answering the fundamental requirements, such as project time and quality, was also defined as the ultimate target (Rouhanizadeh & Kermanshachi, 2019a).

On the other hand, weak consultation with community members, lack of coordination among stakeholders, and lack of voluntary participation have been identified as common barriers in SNDR projects (Kermanshachi & Rouhanizadeh, 2018; Rouhanizadeh et al., 2020). Some scholars believe that the participation of all different stakeholders, from governments, academia, people, and organizations, is necessary for a successful SNDR project (Rehman et al., 2019). However, the shortage of experience in managing people's participation may negatively impact the project and cause delays and cost overruns (Celentano et al., 2019). Furthermore, establishing structures/ guidelines seems vital to avoid such challenges (Flores Salas, 2016). Needless to say, effective communication, which has been regarded as a success factor (Sadiqi et al., 2017), has a prominent role in the whole life cycle of SNDR projects (Yang, 2011); however, the role of communication in emergency management is significant (Han et al., 2019). Effective communication between governments and residents, which can be measured by public anxiety reduction and increment in public trust, can lead to approved planning, improved disaster recovery, especially in time and quality management, better information flow, and enhancement of cohesion among stakeholders (Olshansky, 2006; Rouhanizadeh & Kermanshachi, 2019b; Yang, 2011).

In accordance with the continuous lifespan of SNDR projects, funding and allocating resources should be seen as an ongoing procedure aiming at both short-term and long-term objectives; however, fund allocation should be determined before a disaster, implying the necessity of selecting beneficiaries before the occurrence of catastrophe (Kermanshachi et al., 2019). Moreover, the lack of long-term recovery and funding plans may obstruct the project's success (Dhamangaonkar, 2019). Talking about updated regulations, even though site selection planning should be done before the disaster, many projects postpone site selection until the disaster happens. However, land-use planning can continue to the recovery phase (Edgington, 2011; Ge et al., 2010). Additionally, distinguishing criteria for deciding whether to repair or reconstruct buildings is one of the most demanding topics that need to be considered before a disaster. These regulations should be established before a disaster (Edgington, 2011); however, they will be applied after the disaster. Furthermore, detailed standards to determine retrofit or reconstruction are necessary, and experts should conduct building damage inspections after the disaster (Kermanshachi & Rouhanizadeh, 2018).

3.3 Identifying SNDR Projects' Components by Applying System Thinking

This section presents a new method to comprehensively depict SNDR projects' success factors by applying system thinking. Five sub-systems have been identified, considering SNDR projects as a system (Fig. 3.1). These sub-systems, the main branches of this framework, include the multidimensional aspects influencing SNDR projects. Subsequently, the groups resulting from the division of the sub-systems have been added. The groups work as indicators to determine the influencing factors and place each element appropriately by NVivo 2020. Then, the sub-systems' word clouds representing the highly mentioned topics based on the study's resources have been depicted. Needless to say, the exploit of system thinking, especially in the groups section, enables the researchers to make sure not to leave any influential factors out.

3.3.1 Social Sub-System

As the beneficiaries and users of city facilities, people are the main victims of unsafe conditions, fragile structures, and poor services after a socio-natural disaster. Based on the context analysis, the most frequently mentioned keywords have been shown in Fig. 3.6. "Participation", "community", "public", "social", "people", and "training

Fig. 3.6 The highly mentioned keywords in the social sub-system based on the context analysis

are the ones which have been discussed frequently. Moreover, the literature review has widely mentioned consideration of people's livelihood, safety, culture, and needs.

People's needs can vary; psychological requirements (community empowerment) and on-time answers to residents' fundamental needs, such as food, sanitation, safe shelters, and health services, are part of people's needs (Du et al., 2016; Roosli & O'Keefe, 2013). Comprehensively, the complete restoration of life-line service aims for social recovery (Altay & Green, 2006). Reestablishing community ties that lead to a community willing to work together and revitalize community trust is crucial to forming a sustainable community structure. Providing job opportunities for suffered people has been discussed as the most vital way to the betterment of livelihood (Olshansky, 2006). In addition, long-term payback duration for loans should be designed carefully to prevent people's dependency on outside help (Dunford & Li, 2011).

On the other hand, fair assistance based on people's needs can prevent social and poverty gaps (Zhang & Peacock, 2010). The importance of considering the needs of the people who suffered and their culture in design was studied in numerous papers as it might cause people to refuse recovery outcomes (Tauber, 2015; Cheeck, 2019). Additionally, in the aftermath of a disaster, the security of hit people should be guaranteed (Enshassi et al., 2017). The necessity of criminal behaviour prevention is another missing factor in many disaster recovery projects. On the other side, the difficulty of communication among people and decision-makers is another demanding issue. Improved and effective communication throughout disaster recovery projects leads to better information flow and identifying stakeholders' real needs (Altay & Green,

2006; Rouhanizadeh & Kermanshachi, 2019b). Rising public awareness about safe structures and disaster presentations throughout training, workshops, and meetings should be deemed (Ge et al., 2010; Tauber, 2015). Last but not least, the active engagement of all affected people should be managed (Yilmaz et al., 2013). Based on Fig. 3.9, the researchers divided the social sub-system into "living conditions" and "community participation". The most frequently mentioned factors are "significant level of community participation", "increasing public awareness for the prevention of hazards", "recovery of livelihood", and "effective and holistic communication system".

3.3.2 Economy Sub-System

Although financial management has a significant role in all construction projects, financial challenges might be harsher in the case of SNDR projects due to time limitations and user pressure. Looking at the research references presented in Fig. 3.7, "land", "funding", "economic recovery", "property rights", and "fund allocation" have been discussed widely.

Considering funding as one of the most critical stages in financial management, many disaster recovery projects suffered from insufficient available funds (Johnson, 2006; Li et al., 2019; Saunders et al., 2007). Shortage of financial assistance can affect the project's progress and may negatively impact residents' life (Di Gregorio & Soares, 2017). Regarding the importance of time and resource limitation in SNDR projects, rapid fund collection (Johnson & Lizarralde, 2012) and fair fund allocation (Saunders et al., 2007) are effective keystones that can influence the project's success. On the other hand, in the aftermath of disasters and a considerable level

Fig. 3.7 The highly mentioned keywords in the economy sub-system based on the context analysis

of physical damage to the structures, vague ownership statutes of properties are a burden (Edgington, 2017). Establishing property rights for private and non-private individuals protects people's rights (Johnson & Lizarralde, 2012). The other factor that previous researchers highly stressed was local business recovery (Vahanvati & Mulligan, 2017). Within a revitalized and strengthened business plan for the hit area, the residents' livelihood not only bounced back to stable condition but also some successful cases showed improvement in the public livelihood. Needless to say, the areas' unique characteristics must be considered in the long-term economic development plan (Patel & Hastak, 2013). The economy sub-system has been divided into three groups; "funding for reconstruction", "property rights", and "economy development plan based on area's distinctive features". The most frequently mentioned factors have been identified as "economy development plan based on area's distinctive features", "private & non-private property right protection", and "Sufficient funding availability", as is shown in Fig. 3.9.

3.3.3 Environment Sub-System

In recent years, there has been a rise in the number of studies exploring the sustainable environment and its impacts on human life. This research intends to consider the natural environment as well as the built structures within the environment subsystem. Following the context analysis, "environment", "structures", "buildings", "infrastructures", and "maintenance" have been documented as the highly stressed topics. However, other research concerns are prediction and preparedness, retrofit and repair, secondary disasters, and early warning systems. Figure 3.8 walks the reader through the most frequent topics in the environment sub-system.

On the other hand, recognizing vulnerable buildings before a disaster can be a helpful database for maintenance, retrofitting, and future repair (Du et al., 2016; Saunders et al., 2007). On-time maintenance and retrofitting of the recognized vulnerable buildings, before a disaster or during the reconstruction phase not only can protect people but also save project time (Jigyasu, 2013). The environmental sub-system has two groups; "buildings' conditions" and "natural environment", "recovery of the sustainable environment", "preparedness against natural disaster", "retrofitting vulnerable buildings before a disaster", and "Critical Infrastructures Protection (CIP)" have been appeared more in context analysis (Fig. 3.9).

3.3.4 Government Sub-System

The undeniable role of governments as the main body in charge of SNDR projects varies in different countries; however, the research's findings based on the read resources reveal the most frequently discussed topics. "Coordination", "plans and codes", "financial issues", and "assistance" as well as emergency management,

Fig. 3.8 The highly mentioned keywords in the environment sub-system based on the context analysis

consideration of lessons learnt, and participation management, have been stressed in recent years (Fig. 3.10).

The lack of a strategic management plan causes many disaster recovery projects to suffer, generating pre-existing vulnerabilities, fraud, and corruption (Mukherji, 2008). The well-described and straightforward procedure of the recovery process, including workflow, responsibilities, and protocols (transparency and accountability), and obtainment of building permits alongside the protection of official cooperators' rights such as NGOs, can simplify government regulation and then save time (Du et al., 2016). Even though the role of establishing pre-disaster plans based on the region's historical record of past socio-natural disasters has been studied (Jigyasu, 2013), revision of pre-established plans is crucial due to the dynamic nature of disaster recovery projects (Kermanshachi et al., 2019).

The other prominent subject which is mainly neglected is the identification of different beneficiaries before the disaster. Not only is the recognition of beneficiaries significant, but setting clear guidelines for beneficiary participation by the government is an effective tool in preventing further challenges (Davidson et al., 2007; Griffey, 2010). Government supervision and leadership are vital in all directions of disaster recovery projects (He, 2019). Governments also should manage population density in different areas and control reverse immigration from undamaged regions to hit areas or vice versa (Park & Wang, 2017). The selection and allocation of competent managers and assistants, including NGOs, are also part of the government's responsibility (Yilmaz et al., 2013). Modifying building standards, especially building codes and land use regulations, and the simplified procedure to

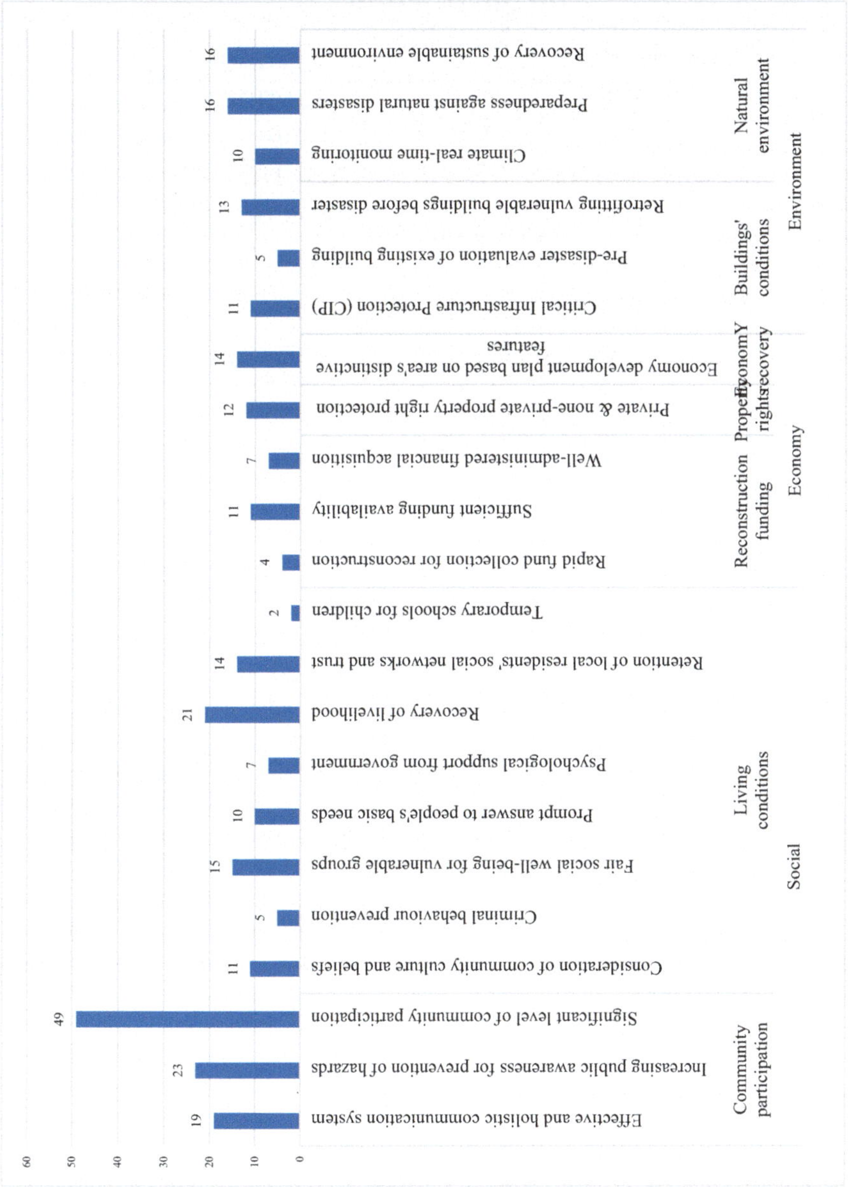

Fig. 3.9 Frequency of factors in the social, economy, and environment sub-systems

Fig. 3.10 The highly mentioned keywords in the government sub-system based on the context analysis

obtain building permits and specific guidelines to determine whether to reconstruct or repair damaged buildings are some of the most important subjects that have been widely discussed (Altay & Green, 2006; He & Zhuang, 2016).

Moreover, insufficient and ill-timed collaboration among in-charge organizations, which has harmful impacts on SNDR projects, has been discussed by some authors (Griffey, 2010; Rehman et al., 2019). Governments are also in charge of financial support (Ding et al., 2019; Du et al., 2016; Siriwardhana et al., 2012). The government sub-system has been divided into three groups; "Policy support", "financial support", and "interactive support mechanism". Figure 3.11 illustrates the most frequently mentioned factors within the three groups. "Up-to-dated reconstruction standards and legislations", "coordination & collaboration among assistants", and "establishment of pre-disaster plans based on past experiences" are three prominent factors in the government sub-system.

3.3.5 *Project Management Sub-System*

Project management is the most crowded sub-system consisting of multidimensional factors. The project's objectives, construction control, planning, and design have been addressed in this sub-system. In accordance with the word cloud in Fig. 3.12, "reconstruction", "stakeholders' needs", "housing", "materials", "resources", and

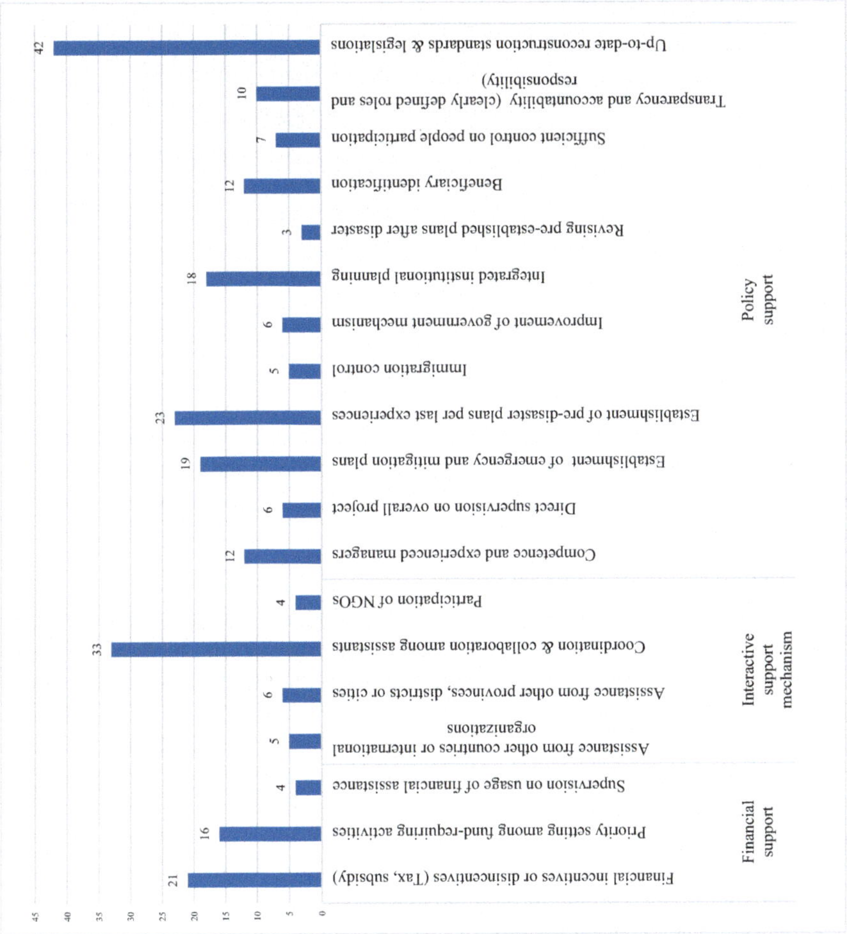

Fig. 3.11 Frequency of factors in the government sub-system

Fig. 3.12 The highly mentioned keywords in the project management sub-system based on the context analysis

"design" have appeared widely in the research literature. However, infrastructures, plans, contractors, technologies, skills, transportation, cost, and location have also gained attention. The study divides the project management sub-system into six groups, as shown in Figs. 3.13 and 3.14.

(1) **Post-Disaster Evaluation**

In the aftermath of a disaster, site investigation and damage assessment are the first necessary steps toe finished successfully to start the recovery process (Ahmed & Charlesworth, 2015). Immediately after disasters, the safety of buildings should be assessed, and further decisions for restoration, repair, or destruction should be made (Xu et al., 2019). A comprehensive and fast damage assessment must cover loss of life, property, infrastructure, livelihoods, and impact on the economy (Moatty, 2017).

(2) **Planning**

Given the rising prevalence of socio-natural disasters and their sudden and destructive effects, conducting comprehensive and detailed planning is essential. Unlike regular construction projects, time and budget limitations and complexity and uncertainty in SNDR projects impose overwhelming pressure on the project team and government (Edgington, 2017; Fayazi et al., 2017; Peng et al., 2021). Focus on short-term activities (linear cycle) has been defined as a topical issue (Supernaw, 2019). Besides, continuous and long-term actions are supposed to bring sustainable outcomes. Moreover, the project's goals and scope should be clearly defined to avoid further misconceptions and conflicts (Hidayat & Egbu, 2010). Although modification of up-to-date standards effectively leads

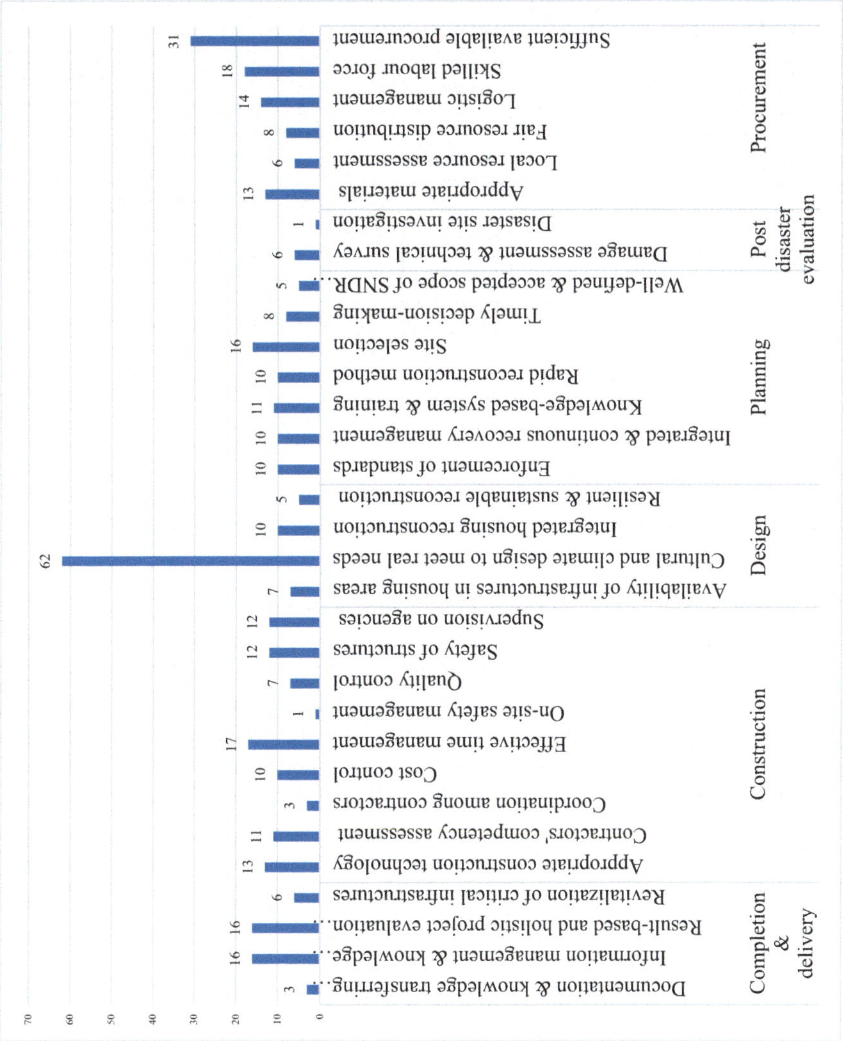

Fig. 3.13 Frequency of factors in the project management sub-system

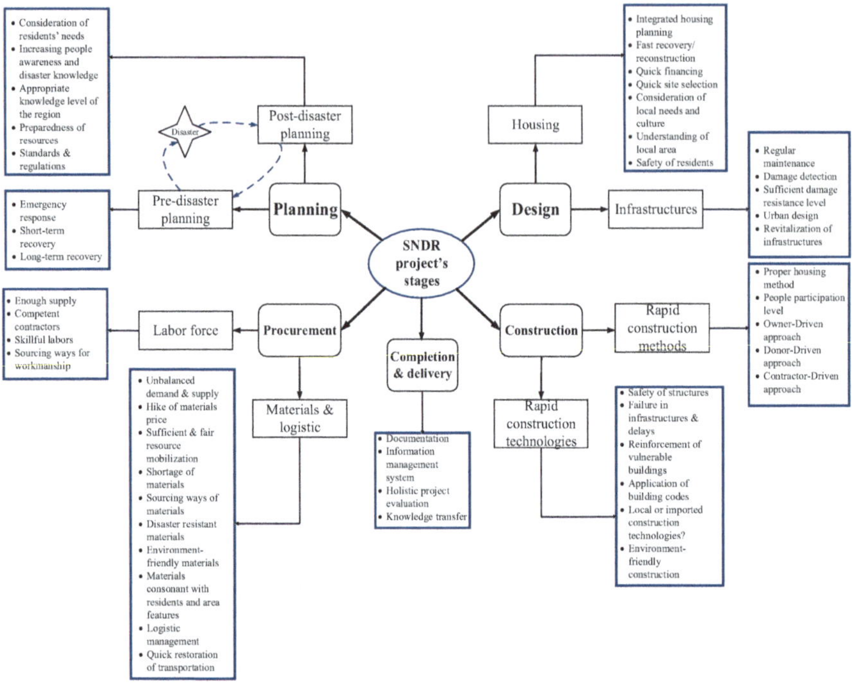

Fig. 3.14 Project management stages of SNDR projects

to SNDR success, without official enforcement, there is still the possibility of standards ignorance by people and implementors (Chen & Booth, 2011). On the other side, the recovery management process should be considered as an integrated management procedure, starting before a disaster and continuing after a disaster occurrence (Mukherji, 2008). Apart from the necessity of enhancement of beneficiaries' knowledge, the matter of time is demanding (Arain, 2015; Zea Escamilla & Habert, 2015). The distance between selected sites and the old site and city facilities is a crucial factor that leads to residents' satisfaction or dissatisfaction (Johnson, 2006). On the other hand, the safety of residents and the prevention of secondary hazards should be considered (Okamoto & Ishikawa, 2013).

(3) **Design**

Assuming the socio-natural disaster recovery project is a system that includes tangible and intangible members, all the features should comply with the system's characteristics. Furthermore, as one of the tangible elements in SNDR projects, buildings should abide by integrated design. To provide a comprehensive understanding of the designing process in SNDR projects, this section explores different attitudes toward designing in "housing" and "infrastructures".

(a) Housing, which has been chiefly suggested to be planned before the disaster (Johnson, 2006), spends almost 30–50% of financial allocation on post-disaster reconstruction projects. The vast investment in the housing sector could be explained by the fact that residents are among the most prominent stakeholders in the SNDR projects (Hidayat & Egbu, 2010). The needs and culture of local residents who have been identified by the government (preferably before the disaster) must be addressed in housing design (Edgington, 2017). Climate design is another vital factor that plays a vital role in people's satisfaction with housing (Tucker et al., 2014). Although standardization in design applied by governments can save time and labour force, this methodology is prone to fail due to differences in climate and culture in regions (Dikmen, 2007). The long-term future of temporary housing should be thought carefully by establishing criteria for deciding about the housing stage (Di Gregorio & Soares, 2017). On the other hand, it is necessary to conduct sustainable design based on the area's previous disasters to avoid the last vulnerabilities (Chang et al., 2011). Additionally, housing reconstruction should be accompanied by the restoration of infrastructures and designing outdoor facilities (Dhamangaonkar, 2019).

(b) Due to the infrastructure's role in revitalizing the whole project, these facilities should bounce back to normalcy as soon as possible (Nejat & Ghosh, 2016). Just after a socio-natural disaster, an insufficient number of operational infrastructures mostly resulted from improper maintenance, which may cause troubles that delays is one of them (Dhamangaonkar, 2019). Several prerequisites need to be achieved for a successful infrastructure designing: effective institutional management, well-defined coordination, well-established regulations, competent managers, effective consultation and communication with stakeholders, easy-to-use information management system, clearly identified goals, and adequate resource management have been identified as the most prominent ones (Ismaila et al., 2014; Liu et al., 2014).

(4) Construction

Based on three primary objectives of construction projects, namely cost, time, and quality, the construction phase needs proper management to balance these goals fairly. Low quality of construction, especially the quality of materials and structures, may generate unsafe buildings, leading to an increment of vulnerabilities (Ahmed & Charlesworth, 2015; Cheng et al., 2017). Shortage of understanding of building standards, non-engineering structures and materials, and lack of supervision during construction are some of the most highly-ranked errors, which can cause low-quality structures in SNDR projects (Chang et al., 2011; Gaillard et al., 2008; Kopaei, 2009; Xu & Lu, 2012; Zhao et al., 2009). Apart from a cost management plan, an effective schedule management plan is needed (Dhamangaonkar, 2019; Sun & Xu, 2011). Quality control mechanisms for construction and material selection and the trade-off between speed and quality can assure pre-established quality standards (Olshansky,

2002). The construction phase has been decomposed into "housing and "infrastructures" from the perspective of "construction methods" and "construction technologies".

(a) While managers are trying to find an appropriate design based on what has been discussed later, selecting a proper housing method should also be contemplated. Each project's selection of an appropriate housing approach depends on cost limitations, the importance of physical and social safety, the revitalization of livelihood, and the community's goals (Enshassi et al., 2017). Housing approaches can generally be divided into three categories: owner-driven reconstruction, donor-driven reconstruction, and contractor-driven reconstruction (Ganapati & Mukherji, 2014; Johnson & Lizarralde, 2012). These approaches are also part of policy strategies known as "Top-down" and "Bottom-up" strategies and have different strengths and weaknesses (Yilmaz et al., 2013).
(b) Although structure failures frequently happen in many construction projects, post-disaster reconstruction projects are more vulnerable to structure failures because of time limitations, affecting structural planning. The selection of proper construction methods depends on the hit area and the local market's capacity to provide labour work and materials, housing design, schedule, and funding availability. Additionally, the possibility of using traditional/vernacular methods or applying new and fast-track construction technologies should be evaluated (Kopaei, 2009; Lizarralde, 2004).

(5) Procurement

As witnessed in reality, many SNDR projects face a lack of resources, which is also mentioned as "supply chain dysfunction" or "inadequate procurement" (Chang et al., 2011; Enshassi et al., 2017; Fayazi et al., 2017). Shortage of reconstruction materials makes disaster recovery projects more susceptible to delays (Ismaila et al., 2014). On the other hand, due to the failure of crucial infrastructures after a disaster, such as roads, accessibility to materials will be affected (Ahmed & Charlesworth, 2015). Furthermore, quick restoration of the transportation system to provide effective logistic management and save time and cost is necessary. To broaden the perspective on sourcing, (Chang et al., 2010a, 2010b) introduced factors affecting the procurement of disaster recovery projects: legislations and policies, the construction industry, the construction market, and the transportation system (Chang et al., 2010a, 2010b). However, this section categorizes procurement into the labour force, materials, and logistics.

(a) The labour force includes both contractors and workers. Indeed, apart from the matter of enough supply of labour force and contractors to expedite recovery speed (Rouhanizadeh & Kermanshachi, 2019b), their competency should also be concentrated (Dhamangaonkar, 2019). Afterwards, cooperation and coordination among contractors, the local government, and other construction agencies must be managed (Meding et al., 2014). Lack of supervision may cause an opaque reconstruction process, illegal reconstructions, poor structural quality, and applied materials (Rouhanizadeh, 2020). Consideration of alternative materials, availability of local materials, and available resource database are three practical solutions offered for solving the shortage of materials and labour force (Silva, 2010).

(b) On the other hand, lack of resources, poor resource management, and doubtful material procurement are the top reasons for SNDR projects' failure (Chang et al., 2010a, 2010b). Reconstruction materials should be chosen based on their resistance against future disasters, high quality, low cost, ability to be used instead of traditional material, and be in compliance with residents' lifestyle and architectural style of the hit area (Ahmed & Charlesworth, 2015; Zhao & Al-Mahaidi, 2009). Transportation and logistics difficulties have been identified as one of the most challenging bottlenecks in post-disaster reconstruction projects (Celentano et al., 2019; Flores Salas, 2016), as it has been proved that transportation has a direct impact on accessibility and price of materials (Chang et al., 2010a, 2010b). Especial attention needs to be paid to a fair distribution of materials among stakeholders over time (Haghani & Afshar, 2010).

(6) **Completion & Delivery**

Establishing an information dissemination system and keeping stakeholders informed about the SNDR project will lead to faster and preferable recovery (UNISDR, 2015). Additionally, insufficient knowledge among team members requires upgrading knowledge transfer mechanisms (Thanurjan & Seneviratne, 2009). Documentation protocol alongside information management systems may deepen beneficiaries' understanding of socio-natural disasters, especially in prevention and preparation (Ahmed & Charlesworth, 2015). From the perspective of upper-level managers and decision-makers, an information and documentation dashboard will be intensely worthwhile to lessen disasters' destructive effects and plan timely responses (Vahanvati & Mulligan, 2017). Additionally, the majority of disaster recovery projects were assessed based on physical results such as the number of built structures and completion duration (Robert et al., 2005). However, the assessment of SNDR projects as complex systems requires a multi-factor index system to cover all the influential factors (Akiyama, 2019; Supernaw, 2019).

3.4 Conceptual Project Management Framework of SNDR Projects

Since the project management sub-system owns many groups and factors, this chapter attempts to sort out the prevalent and most mentioned activities considering the project management sub-system. Therefore, proposing a continuous lifespan

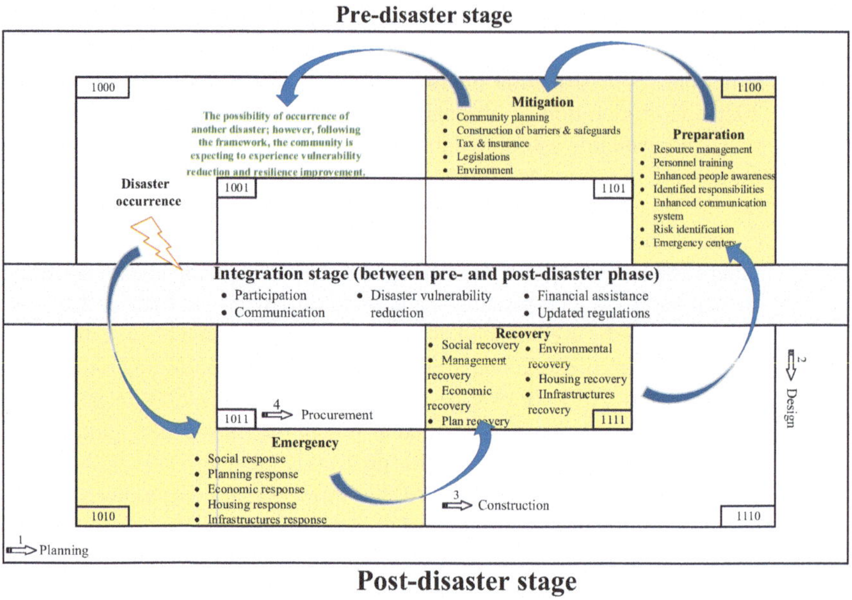

Fig. 3.15 Conceptual framework for the management of SNDR projects

of SNDR projects and establishing project stages have been identified as practical solutions. The project management framework, presented in Sect. 3.4., results from applying system thinking and life-cycle decomposition for the project management sub-system. The combination of project management stages, SNDR projects, and the life cycle is shown in Fig. 3.15. Based on the available data, it can be proposed that:

(1) According to the definition of a project by IPMI, "*a project is a temporary endeavour undertaken to create a unique product or service. Temporary means that every project has a definite beginning and a definite end*" (Institute, 2000). Hence, it is necessary to clarify the ending and beginning of SNDR projects because they differ from regular projects. Bouncing back to what has been destroyed in a glance may take several years. Furthermore, short-term recovery plans cannot address community needs. The concept of Build Back Better, which has been focused on in recent years, also points to the necessity of establishing long-term plans. Moreover, the traditional management model for recovery projects has a linear process, prone to methodologic inadequacies due to the fragmented phases emphasizing outcomes rather than process (Moatty, 2017). This section suggested a conceptual framework attempting to combine lifespan for SNDR projects and project management stages. As the results surpass the earlier work in this area regarding the necessity of linking the development recovery to the mitigation and preparedness phases, a non-closed loop cycle has been utilized, representing the lifespan of the recovery projects.

(2) The identified project steps based on the definition suggested by the IPMI and stages of a post-disaster recovery project by Silva (2010) have been combined with the disaster recovery lifespan (Silva, 2010). A comprehensive management framework covers the recovery of structures and community well-being, and improvement in social and economic aspects should be addressed. A Venn diagram then assisted the study in showing the interrelationships between the project management stages and the suggested lifespan. Planning is always present as the first stage of project management, while design, construction, and procurement might be absent in some phases of the project lifespan. The Venn diagram enjoys four-digit codes for each area of the rectangles to show the presence or absence of "planning", "design", "construction", and "procurement". By use of three rectangles. While zero stands for the absence of the stages, one indicates the presence of the stages. The presence of planning has been shown by the inside areas of the biggest rectangle. Planning owns the first digit number in the codes. The middle rectangle, representing the design (the second digit number) phase on the horizontal line and the construction stage (the second digit number) on the vertical line, divides the space into four areas. Finally, the fourth digit number presents the procurement stage, and its presence is shown in the areas located inside the smallest rectangle. In the course of this work, it has been revealed that planning is the most paramount step in all phases of disaster recovery projects.

(3) The suggested framework also reveals the necessity of planning before disasters occur. This observation also agrees with the results reported by Bosher et al. (2021), as they have demonstrated the need to pay more attention to disaster mitigation and preparedness rather than emergency management (Bosher et al., 2021). Furthermore, the more resources are being used in the mitigation and preparedness stage, the more sustainable the communities will be.

(4) This framework reveals that some activities cannot be placed in just one stage of the SNDR life cycle. These activities, which should be followed during the project's entire life cycle, have been grouped in the "integration phase", an elastic stage. Apparently, stakeholders' participation in the SNDR project can both happen in the mitigation and preparedness phases, while nobody can ignore the role of beneficiary involvement in the recovery stage. An effective communication system is also needed for the entire SNDR project.

(5) Interestingly, the recovery phase seems to be the most crowded stage based on the project management steps. This stage includes all the steps of the research's proposed classification of project management, namely planning, design, construction, and procurement. Restoration or reconstruction of damaged houses and infrastructures is supposed to be done here. Therefore, it is not unexpected to see that post-disaster reconstruction is just one part of the more significant concept of "socio-natural disaster recovery".

(6) In the process of proposing the conceptual framework for the management of the SNDR project, a great deal of understanding was gained on the tasks that might affect the projects. The framework provides a solid foundation on which future research can be built to manage SNDR projects properly. The combination of

project management lifespan and project management stages has endorsed the necessity of having a continuous management plan that must be declared clearly. This chapter's results will lead to identifying the success dimensions of SNDR projects.

3.5 Chapter Summary

In summary, the main contributions of this chapter can be considered as follows:

- System thinking has significantly simplified the content of disaster recovery projects through its ability to see the system as a whole while manifesting the interrelations among the components.
- On the other hand, the proposed conceptual management framework for the SNDR project attempts to identify and locate the most significant recovery activities considering the continual SNDR project life cycle. The successive stages in this framework are expected to solve the loss of the experience and knowledge gained throughout SNDR projects to be ready for imminent socio-natural disasters.
- Although revitalizing the destroyed facilities and buildings greatly impacts the recovery of other damaged aspects, negligence in the recovery of the intangible sides, such as the economy and environment, may cause failure to reach disaster recovery goals. A comprehensive recovery can be followed by applying the research's proposed framework. Additionally, the significance of planning has been proved. Planning is the main task that should be seen in the entire SNDR project's life cycle.
- By dividing the lifespan of the SNDR project into the pre-disaster stage and post-disaster stage, some activities have been grouped in the "integration stage", which should be followed during the entire SNDR project.
- Finally, the results illustrate that the recovery phase has been identified as the most crowded stage, considering the number of assigned project stages. Post-disaster reconstruction is part of the more significant cycle named socio-natural disaster recovery.

References

Ahmed, I., & Charlesworth, E. R. (2015). An evaluation framework for assessing resilience of post-disaster housing. *International Journal of Disaster Resilience in the Built Environment, 6*(3), 300–312. https://doi.org/10.1108/IJDRBE-11-2013-0042

Akiyama, H. (2019). *Who rebuilds? Civil society in post-disaster recovery in China and Japan.* George Mason University.

Altay, N., & Green, W. G. (2006). Interfaces with other disciplines OR / MS research in disaster operations management. *European Journal of Operational Research, 175*(December 2004), 475–493. https://doi.org/10.1016/j.ejor.2005.05.016

Arain, F. (2015). Knowledge-based approach for sustainable disaster management: Empowering emergency response management team. *Procedia Engineering, 118*(2015), 232–239. https://doi.org/10.1016/j.proeng.2015.08.422

Arnold, R. D., & Wade, J. P. (2015). A definition of systems thinking: A systems approach. *Procedia Computer Science, 44*, 669–678. https://doi.org/10.1016/j.procs.2015.03.050

Bosher, L., Chmutina, K., & van Niekerk, D. (2021). Stop going around in circles: Towards a reconceptualisation of disaster risk management phases. *Disaster Prevention and Management: An International Journal, 30*(4/5), 525–537. https://doi.org/10.1108/DPM-03-2021-0071

Cabrera, D., Colosi, L., & Lobdell, C. (2008). Systems thinking. *Evaluation and Program Planning, 31*(3), 299–310. https://doi.org/10.1016/j.evalprogplan.2007.12.001

Celentano, G., Escamilla, E. Z., Göswein, V., & Habert, G. (2019). A matter of speed: The impact of material choice in post-disaster reconstruction. *International Journal of Disaster Risk Reduction, 34*(2019), 34–44. https://doi.org/10.1016/j.ijdrr.2018.10.026

Chang, Y., Wilkinson, S., Potangaroa, R., & Seville, E. (2010). Resourcing challenges for post-disaster housing reconstruction: A comparative analysis. *Building Research and Information, 38*(3), 247–264. https://doi.org/10.1080/09613211003693945

Chang, Y., Wilkinson, S., Potangaroa, R., & Seville, E. (2011). Donor-driven resource procurement for post-disaster reconstruction: Constraints and actions. *Habitat International, 35*(2), 199–205. https://doi.org/10.1016/j.habitatint.2010.08.003

Chang, Y., Wilkinson, S., Seville, E., & Potangaroa, R. (2010b). Resourcing for a resilient post-disaster reconstruction environment. *International Journal of Disaster Resilience in the Built Environment, 1*(1), 65–83. https://doi.org/10.1108/17595901011026481

Cheeck, W. W. (2019). The paradox of community involvement: Post-disaster reconstruction in Minamisanriku Japan. In *ProQuest dissertations and theses*. Tulane University.

Chen, Y., & Booth, D. C. (2011). The Wenchuan earthquake of 2008 anatomy of a disaster. In *Zhurnal Eksperimental'noi i Teoreticheskoi Fiziki*. Springer. https://doi.org/10.1007/978-3-642-21159-1

Cheng, M. Y., Chiu, K. C., Hsieh, Y. M., Yang, I. T., Chou, J. S., & Wu, Y. W. (2017). BIM integrated smart monitoring technique for building fire prevention and disaster relief. *Automation in Construction, 84*(2017), 14–30. https://doi.org/10.1016/j.autcon.2017.08.027

Da Silva, J. (2010). Lessons from Aceh. Key considerations in post-disaster reconstruction. In *Post-disaster reconstruction: Lessons from Aceh*. Practical Action Publishing. https://doi.org/10.4324/9781849775137

Davidson, C. H., Johnson, C., Lizarralde, G., Dikmen, N., & Sliwinski, A. (2007). Truths and myths about community participation in post-disaster housing projects. *Habitat International, 31*(2007), 100–115. https://doi.org/10.1016/j.habitatint.2006.08.003

Dhamangaonkar, V. (2019). Identification of legal and policy barriers to timely post-disaster reconstruction and strategy development for effective recovery processes in United States [The University of Texas]. In *ProQuest dissertations and theses* (Issue May). https://doi.org/.1037//0033-2909.I26.1.78

Di Gregorio, L. T., & Soares, C. A. P. (2017). Post-disaster housing recovery guidelines for development countries based on experiences in the American continent. *International Journal of Disaster Risk Reduction, 24*(2017), 340–347. https://doi.org/10.1016/j.ijdrr.2017.06.027

Dikmen, N. (2007). Relocation or rebuilding in the same area: An important factor for decision making for post disaster housing projects. In D. Alexander, C. H. Davidson, A. Fox, G. Lizzaralde & C. Johnson (Eds.), *Proceedings of the International Conference and Student Competition on Post-disaster Reconstruction - Meeting Stakeholder Interests*. Firenze University Press.

Ding, Z., Liu, S., Liao, L., & Zhang, L. (2019). A digital construction framework integrating building information modeling and reverse engineering technologies for renovation projects. *Automation in Construction, 102*(2019), 45–58. https://doi.org/10.1016/j.autcon.2019.02.012

Du, F., Okazaki, K., Ochiai, C., & Kobayashi, H. (2016). Post-disaster building repair and retrofit in a disaster-prone historical village in China: A case study in Shangli, Sichuan. *International Journal of Disaster Risk Reduction, 16*(2016), 142–157. https://doi.org/10.1016/j.ijdrr.2016.02.007

Dunford, M., & Li, L. (2011). Earthquake reconstruction in Wenchuan: Assessing the state overall plan and addressing the "forgotten phase." *Applied Geography, 31*(3), 998–1009. https://doi.org/10.1016/j.apgeog.2011.01.001

Edgington, D. W. (2011). Reconstruction after natural disasters: The opportunities and constraints facing our cities. In *Town planning review* (Vol. 82, Issue 6). https://doi.org/10.3828/tpr.2011.35

Edgington, D. W. (2017). Building back better along the Sanriku coast of Tohoku, Japan: Five years after the "3/11" disaster. *Town Planning Review, 88*(6), 615–638. https://doi.org/10.3828/tpr.2017.38

Enshassi, A., Chatat, T., von Meding, J., & Forino, G. (2017). Factors influencing post-disaster reconstruction project management for housing provision in the Gaza Strip, occupied Palestinian territories. *International Journal of Disaster Risk Science, 8*(4), 402–414. https://doi.org/10.1007/s13753-017-0155-4

Fan, X., Juang, C. H., Wasowski, J., Huang, R., Xu, Q., Scaringi, G., van Westen, C. J., & Havenith, H. B. (2018). What we have learned from the 2008 Wenchuan Earthquake and its aftermath: A decade of research and challenges. *Engineering Geology, 241*(July 2018), 25–32. https://doi.org/10.1016/j.enggeo.2018.05.004

Fan, Y., Li, Z., Pei, J., Li, H., & Sun, J. (2015). Applying systems thinking approach to accident analysis in China: Case study of "7.23" Yong-Tai-Wen high-speed train accident. *Safety Science, 76*(2015), 190–201. https://doi.org/10.1016/j.ssci.2015.02.017

Fayazi, M., Arefian, F. F., Gharaati, M., Johnson, C., Lizarralde, G., & Davidson, C. (2017). Managing institutional fragmentation and time compression in post-disaster reconstruction – the case of Bam. *International Journal of Disaster Risk Reduction, 21*(September 2016), 340–349. https://doi.org/10.1016/j.ijdrr.2017.01.012

Flores Salas, A. (2016). Evaluation of the use of lightweight concrete panels for post disaster house reconstruction using building information modelling. In *PQDT - UK & Ireland*. University of Manchester.

Gaillard, J. C., Clavé, E., & Kelman, I. (2008). Wave of peace? Tsunami disaster diplomacy in Aceh, Indonesia Jean-Christophe. *Geoforum, 39*(1), 511–526. https://doi.org/10.1016/j.geoforum.2007.10.010

Ganapati, N. E., & Mukherji, A. (2014). Out of sync: World bank funding for housing recovery, postdisaster planning, and participation. *Natural Hazards Review, 15*(1), 58–73. https://doi.org/10.1061/(asce)nh.1527-6996.0000120

Ge, Y., Gu, Y., & Deng, W. (2010). Evaluating China's national post-disaster plans: The 2008 Wenchuan earthquake's recovery and reconstruction planning. *International Journal of Disaster Risk Science, 1*(2), 17–27. https://doi.org/10.3974/j.issn.2095-0055.2010.02.003

Ghaffarian, S., Farhadabad, A. R., & Kerle, N. (2020). Post-disaster recovery monitoring with google earth engine. *Applied Sciences (Switzerland), 10*(13), 4574. https://doi.org/10.3390/app10134574

Ghaffarian, S., Roy, D., Filatova, T., & Kerle, N. (2021). Agent-based modelling of post-disaster recovery with remote sensing data. *International Journal of Disaster Risk Reduction, 60*(2021), Article 102285. https://doi.org/10.1016/j.ijdrr.2021.102285

Griffey, A. M. (2010). *Infrastructure disaster management: Insight from small business leaders after a mega disaster* (Issue December). University of Phoenix.

Guo, Y. (2012). Urban resilience in post-disaster reconstruction: Towards a resilient development in Sichuan, China. *International Journal of Disaster Risk Science, 3*(1), 45–55. https://doi.org/10.1007/s13753-012-0006-2

Haghani, A., & Afshar, A. M. (2010). *Development of advanced applications using bluetooth-generated traffic flow data* (Issue January).

Han, J., Zhang, P., & Song, Y. (2019). The construction of emergency communication model based on the Kite-type co-orientation approach. In H. Tan (Ed.), *Technology for education and learning* (Vol. 53, Issue 9, pp. 235–243). Springer. https://doi.org/10.1017/CBO9781107415324.004

Han, Z. (2014). *From vulnerability to resilience: Long-term livelihood recovery in rural china after the 2008 Wenchuan earthquake*. University of Delaware.

He, F., & Zhuang, J. (2016). Balancing pre-disaster preparedness and post-disaster relief. *European Journal of Operational Research, 252*(1), 246–256. https://doi.org/10.1016/j.ejor.2015.12.048

He, L. (2019). Identifying local needs for post-disaster recovery in Nepal. *World Development, 118*(2019), 52–62. https://doi.org/10.1016/j.worlddev.2019.02.005

He, L., Dominey-Howes, D., Aitchison, J. C., Lau, A., & Conradson, D. (2021). How do post-disaster policies influence household-level recovery? A case study of the 2010–11 Canterbury earthquake sequence, New Zealand. *International Journal of Disaster Risk Reduction, 60*(August 2020), 102274. https://doi.org/10.1016/j.ijdrr.2021.102274

Hidayat, B., & Egbu, C. (2010). A literature review of the role of project management in post-disaster reconstruction. In *Procs 26th Annual* ARCOM *Conference* (pp. 1269–1278). September.

Institute, P. M. (2000). PMBOK® guide. In *Project management institute, Inc.* (2000th ed.). https://doi.org/10.1093/ajcp/69.5.475

Ismaila, D., Majidb, T. A., Rooslic, R., & Samahd, N. A. (2014). Project management success for post-disaster reconstruction projects: International NGOs perspectives. *Procedia Economics and Finance, 18*(2014), 120–127. https://doi.org/10.1016/s2212-5671(14)00921-6

Jigyasu, R. (2013). Long-term cultural impacts of disaster decision-making: The case of post earthquake reconstruction in Marathwada, India. *International Journal of Architectural Research, 7*(3), 14–23. https://doi.org/10.26687/archnet-ijar.v7i3.46

Johnson, C. (2006). *Strategic planning* for *temporary housing: 1999 earthquakes in Turkey*. University of Montreal.

Johnson, C., & Lizarralde, G. (2012). Post-disaster housing and reconstruction. In *International encyclopedia of housing and home* (pp. 340–346). https://doi.org/10.1016/B978-0-08-047163-1.00046-1

Jordan, E. (2012). *Pathways to community recovery: A qualitative comparative analysis of post-disaster outcomes*. University of Colorado.

Jordan, E., Javernick-Will, A., & Tierney, K. (2016). Post-tsunami recovery in Tamil Nadu, India: Combined social and infrastructural outcomes. *Natural Hazards, 84*(2), 1327–1347. https://doi.org/10.1007/s11069-016-2489-4

Kermanshachi, S., Bergstrand, K., & Rouhanizadeh, B. (2019). *Identifying, weighting and causality modeling of social and economic barriers to rapid infrastructure recovery from natural disasters: A study of Hurricanes Harvey, Irma and Maria.*

Kermanshachi, S., & Rouhanizadeh, B. (2018). Feasibility analysis of post disaster reconstruction alternatives using automated BIM-based construction cost estimation tool. In *Building tomorrow's society* (pp. 1–10).

Kopaei, M. G. (2009). Knowledge transfer in post-disaster reconstruction: The problem of post-post-disaster reconstruction. In *ProQuest dissertations and theses* (Issue May). McGill University.

Li, Q., Umaier, K., & Koide, O. (2019). Research on post-Wenchuan earthquake recovery and reconstruction implementation: A case study of housing reconstruction of Dujiangyan City. *Progress in Disaster Science, 4*, 10041. https://doi.org/10.1016/j.pdisas.2019.100041

Liu, L., Lin, Y., & Wang, S. (2014). Urban design for post-earthquake reconstruction: A case study of Wenchuan county, China. *Habitat International, 41*, 290–299. https://doi.org/10.1016/j.habitatint.2013.09.001

Liu, Z., & Han, Y. (2009). Post-disaster reconstruction and social work development in the Chinese mainland. *China Journal of Social Work, 2*(3), 221–223. https://doi.org/10.1080/17525090903211796

Lizarralde, G. (2004). *Organisational system and performance of post-disaster reconstruction projects*. Universite de Montreal.

Lu, Y., & Xu, J. (2015). Comparative study on the key issues of postearthquake recovery and reconstruction planning: Lessons from the United States, Japan, Iran, and China'. *Natural Hazards Review, 16*(3), 04014033. https://doi.org/10.1061/(asce)nh.1527-6996.0000172

Moatty, A. (2017). Post-flood recovery: An opportunity for disaster risk reduction? In F. Vinet (Ed.), *Floods* (Vol. 2, pp. 350–363). Elsevier. https://doi.org/10.1016/B978-1-78548-269-4.50023-8

Mukherji, A. (2008). Negotiating housing recovery: Why some communities recovered while others struggled to rebuild in post-earthquake urban Kutch, India. In *Curr. Biol.* (Vol. 11, Issue 15). University of California.

Nejat, A., & Ghosh, S. (2016). LASSO model of postdisaster housing recovery: Case study of hurricane sandy. *Natural Hazards Review, 17*(3), 04016007. https://doi.org/10.1061/(asce)nh.1527-6996.0000223

Okamoto, K., & Ishikawa, Y. (2013). Traditional wisdom and modern knowledge for the earth's future. In J. Noritaka Yagasaki (Ed.), *International Geographical Union Kyoto Regional Conference*. Springer.

Olshansky, R. (2002). Planning for disasters. *Journal of the American Planning Association, 68*(4), 453.

Olshansky, R. B. (2006). Planning after Hurricane Katrina. *Journal of the American Planning Association, 72*(2), 147–153. https://doi.org/10.1080/01944360608976735

Park, A., & Wang, S. (2017). Benefiting from disaster public and private responses to the Wenchuan earthquake. *World Development, 94*, 38–50. https://doi.org/10.1016/j.worlddev.2016.12.038

Patel, S., & Hastak, M. (2013). A framework to construct post-disaster housing. *International Journal of Disaster Resilience in the Built Environment, 4*(1), 95–114. https://doi.org/10.1108/17595901311299026

Peng, L., Li, Q., Deng, W., & Liu, Y. (2021). What promotes post-earthquake economic recovery: The role of counterpart assistance policy after the Wenchuan ms 8.0 earthquake, China. *SAGE Open, 11*(3), 1–14. https://doi.org/10.1177/21582440211033573

Platt, S. (2018). Factors affecting the speed and quality of post-disaster recovery and resilience. In S. Ó. Rajesh Rupakhety (Ed.), *Earthquake Engineering and structural dynamics in memory of Ragnar Sigbjörnsson* (pp. 369–403). Springer. https://doi.org/10.1007/978-3-319-62099-2

Rani, W. N. M. W. M., Nifa, F. A. A., Ismail, M. N., & Khalid, K. N. (2017). Planning for post disaster recovery: Lesson learnt from flood events in Kelantan Malaysia. In *AIP Conference Proceedings* (Vol. 1891, No. 1, p. 020143). https://doi.org/10.1063/1.5005476

Rehman, J., Sohaib, O., Asif, M., & Pradhan, B. (2019). Applying systems thinking to flood disaster management for a sustainable development. *International Journal of Disaster Risk Reduction, 36*(February), 101101. https://doi.org/10.1016/j.ijdrr.2019.101101

Robert, O., Ikuo, K., & Kazuyoshi, O. (2005). The Kobe earthquake, ten years later. *Planning, 71*(9), 36.

Rodrigo, N., & Wilkinson, S. (2021). Impact of flagship projects on the recovery of a city post-disaster. *International Journal of Disaster Risk Reduction, 58*(February), 102191. https://doi.org/10.1016/j.ijdrr.2021.102191

Roosli, R., & O'Keefe, P. (2013). Post-disaster housing and management in Malaysia: A literature review. In *International journal of disaster resilience in the built environment* (Vol. 4, Issue 2, pp. 168–181). https://doi.org/10.1108/IJDRBE-06-2011-0022

Rose, A. (2011). Resilience and sustainability in the face of disasters. *Environmental Innovation and Societal Transitions, 1*(2011), 96–100. https://doi.org/10.1016/j.eist.2011.04.003

Rouhanizadeh, B., & Kermanshachi, S. (2019a). Comparative analysis of public's and decision-maker's perspectives on socioeconomic barriers causing delay in post-disaster recovery processes. In *ASCE Construction Research Congress (CRC)*, September, 1–10. https://doi.org/10.1061/9780784482858.092

Rouhanizadeh, B., & Kermanshachi, S. (2019b). Investigating the relationships of socioeconomic factors delaying post-disaster reconstruction. In C. Wang, Y. K. Cho, F. Leite & A. Behzadan (Ed.), *ASCE International Conference on Computing in Civil Engineering* (pp. 16–24). https://doi.org/10.6224/JN.59.5.16

Rouhanizadeh, B., Kermanshchi, S., & Dhamangaonkar, V. S. (2020). Reconstruction of critical and interdependent infrastructures due to catastrophic natural disasters: Lessons learned. In D. Grau, M. E. Asmar & P. Tang (Eds.), *Construction Research Congress 2020*. ASCE.

Sadiqi, Z., Trigunarsyah, B., & Coffey, V. (2017). A framework for community participation in post-disaster housing reconstruction projects: A case of Afghanistan. *International Journal of Project Management, 35*(5), 900–912. https://doi.org/10.1016/j.ijproman.2016.11.008

Saunders, M., Lewis, P., & Thornhill, A. (2007). Research methods for business students. In *Pearson education limited 2.*

Supernaw, S. A. (2019). *Impact of project management methodology use in post-disaster recovery and mitigation projects by North Carolina emergency managers* (Issue February). Northcentral University.

Sheykhmousa, M., Kerle, N., Kuffer, M., & Ghaffarian, S. (2019). Post-disaster recovery assessment with machine learning-derived land cover and land use information. *Remote Sensing, 11*(10), 1174. https://doi.org/10.3390/rs11101174

Siriwardhana, C., Hewage, S., Deshabandu, R., Siribaddana, S., & Sumathipala, A. (2012). Psychosocial and ethical response to disasters: A SWOT analysis of post-tsunami disaster management in Sri Lanka. In *Asian bioethics review* (Vol. 4, Issue 3, pp. 171–182). https://doi.org/10.1353/asb.2012.0027

Sun, C., & Xu, J. (2011). Estimation of time for Wenchuan earthquake reconstruction in China. *Journal of Construction Engineering and Management, 137*(3), 179–187. https://doi.org/10.1061/(ASCE)CO.1943-7862.0000277

Tauber, G. (2015). Architects and rural post-disaster housing: Lessons from south india. In *International journal of disaster resilience in the built environment* (Vol. 6, Issue 2, pp. 206–224). https://doi.org/10.1108/IJDRBE-07-2013-0025

Thanurjan, R., & Seneviratne, L. D. I. P. (2009). The role of knowledge management in post-disaster housing reconstruction. *Disaster Prevention and Management: An International Journal, 18*(1), 66–77. https://doi.org/10.1108/09653560910938556

Tucker, S., Gamage, A., & Wijeyesekera, C. (2014). Some design aspects of sustainable post-disaster housing. *International Journal of Disaster Resilience in the Built Environment, 5*(2), 163–181. https://doi.org/10.1108/IJDRBE-06-2012-0019

UNISDR. (2015). *Considerations on developing a system of indicators based on the sendai framework for disaster risk reduction 2015–2030 : A proposal for monitoring progress.*

Vahanvati, M., & Mulligan, M. (2017). A new model for effective post-disaster housing reconstruction: Lessons from Gujarat and Bihar in India. *International Journal of Project Management, 35*(5), 802–817. https://doi.org/10.1016/j.ijproman.2017.02.002

von Meding, J., Oyedele, L., & Bruen, J. (2014). Linking organisational competency to project success in post-disaster reconstruction. *Open House International, 39*(3), 7–16.

Walker, B., de Vries, H. P., & Nilakant, V. (2017). Managing legitimacy: The Christchurch post-disaster reconstruction. *International Journal of Project Management, 35*(5), 853–863. https://doi.org/10.1016/j.ijproman.2016.07.007

Wu, J. Y. (2003). A comparative study of housing reconstruction after two major earthquakes: The 1994 Northridge earthquake in the United States and the 1999 Chi-Chi earthquake in Taiwan. In *ProQuest dissertations and theses* (Vol. 28, Issue August). Texas A&M University.

Xiao, Y., Olshansky, R., Zhang, Y., Johnson, L. A., & Song, Y. (2020). Financing rapid community reconstruction after catastrophic disaster: Lessons from the 2008 Wenchuan earthquake in China. *Natural Hazards, 104*(s1), 5–30. https://doi.org/10.1007/s11069-019-03789-9

Xu, H., Wang, L., Li, Y., Huang, L., & Yin, L. (2019). Critical Chinese government behavior for swift housing reconstruction programme - The case of Dujiangyan following Wenchuan earthquake. *International Journal of Disaster Risk Reduction, 42*, Article 101307. https://doi.org/10.1016/j.ijdrr.2019.101307

Xu, J., & Lu, Y. (2012). Meta-synthesis pattern of post-disaster recovery and reconstruction: Based on actual investigation on 2008 Wenchuan earthquake. *Natural Hazards, 60*(2), 199–222. https://doi.org/10.1007/s11069-011-0003-6

Xu, J., & Lu, Y. (2018). Towards an earthquake-resilient world: From post-disaster reconstruction to pre-disaster prevention. In *Environmental Hazards* (Vol. 17, Issue 4, pp. 269–275). https://doi.org/10.1080/17477891.2018.1500878

Xu, P., Lu, X., Zuo, K., & Zhang, H. (2014). Post-Wenchuan earthquake reconstruction and development in China. In K. T. L. Naim Kapucu (Ed.), *Disaster and development examining global issues and cases*. Springer. https://doi.org/10.1007/978-3-319-04468-2_25

Yang, B. (2011). *A comprehensive study of the Chinese government's natural disaster communication framework based on the natural disaster management and communication system's response to two natural disasters* (Issue August). University of Southern California.

Yilmaz, D. G., Meding, J. V., & Erk, G. K. (2013). A theoretical approach to the design of a survey instrument in post-disaster reconstruction: Defining indicators for a human-based study in rural built-environment. *International Journal of Architectural Research, 7*(3), 40–56.

Zea Escamilla, E., & Habert, G. (2015). Global or local construction materials for post-disaster reconstruction? Sustainability assessment of 20 post-disaster shelter designs. *Building and Environment, 92*(2015), 692–702. https://doi.org/10.1016/j.buildenv.2015.05.036

Zhang, W., Zhu, S., Zhang, X., & Zhao, T. (2020). Identification of critical causes of construction accidents in China using a model based on system thinking and case analysis. *Safety Science, 121*(April 2019), 606–618. https://doi.org/10.1016/j.ssci.2019.04.038

Zhang, Y., & Peacock, W. G. (2010). Planning for housing recovery? Lessons learned from Hurricane Andrew. *Journal of the American Planning Association, 76*(1), 5–24. https://doi.org/10.1080/01944360903294556

Zhao, B., Taucer, F., & Rossetto, T. (2009). Field investigation on the performance of building structures during the 12 May 2008 Wenchuan earthquake in China. *Engineering Structures, 31*(8), 1707–1723. https://doi.org/10.1016/j.engstruct.2009.02.039

Zhao, X. L., & Al-Mahaidi, R. (2009). Wenchuan post-earthquake reconstruction by utilizing composite Tubluar construction and FRP retrofitting technology. *Journal of Sichuan University (Engineering Science Edition), 41*(3), 231–247.

Chapter 4
Structuring the Framework of Primary and Selected Success Factors Affecting Socio-Natural Disaster Recovery Projects

Abstract SNDR projects are multidimensional subjects whose success relies on the progress of sub-groups (Ganapati & Mukherji in JAMA 15(1):58–73, 2014). Although the number of studies exploring the reasons for failure of disaster recovery projects is rising, they generally concentrate on recovery projects as a single short-term reconstruction task. Furthermore, to propose a set criterion for evaluating SNDR projects' recovery, an appropriate answer should be found for the following question: *Which factors can be holistically influential on the recovery projects after a socio-natural disaster?* Selected Success Factors (SSFs) are the necessary elements to form CSFs, which are causes of project success. Once the CSFs have been developed, Key Performance Indicators (KPIs) can be structured to evaluate the success of SNDR projects (Yilmaz et al., in JAMA 7(3):40–56, 2013). CSFs have been traditionally classified according to time, cost, quality, and stakeholders' desire (Rani et al., in JAMA 1891(1), 2017; Supernaw in Impact of project management methodology use in post-disaster recovery and mitigation projects by North Carolina emergency managers, Northcentral University, 2019); however, some modifications are necessary to make the evaluation criteria appropriate to evaluate SNDR projects. Furthermore, based on the research outcomes in Fig. 4.2, this chapter attempted to (a) identify Primary Success Factors (PSFs) using system thinking and (b) structure SSFs by reliability check of the PSFs per experts' judgments. Moreover, the results gained from the Delphi survey and critical literature review were compared to identify the difference between experts' and literature perspectives.

Keywords Successful disaster recovery project · Delphi survey · Success factors

4.1 Primary Success Factors in Disaster Recovery Projects

A growing body of literature discusses socio-natural disaster recovery projects' life cycle. The importance of "pre-disaster planning" and "post-disaster planning" has been focused on by many scholars (Kermanshachi et al., 2019; Xu & Lu, 2018). Present-day science indicates that SNDR projects have a complex nature, requiring

H. Bahmani et al., *Socio-Natural Disaster Recovery Projects*, Resilient and Friendly Living Environments, https://doi.org/10.1007/978-981-95-0113-7_4

balancing pre-disaster preparedness and post-disaster relief. The necessity of continuity of the pre-disaster activities to the recovery phase and considering an integration phase to use the disaster's window of opportunity to flourish potential chances was stressed in some studies (Wu, 2003). Learning from the past, which has been called the "*Proactive approach*", may impact pre-disaster planning and cause quick response after the disaster, which can save time (Griffey, 2010; Olshansky, 2002). Within this lens, several studies proposed a model for the pre-diaster phases (Chen & Booth, 2011; Yilmaz et al., 2013). Another comprehensive life cycle of SNDR projects covers various topics from the beginning of the disaster to the development phase (Dunford & Li, 2011b). Figure 4.1 represents a diagram to understand socio-natural disasters' continuous recovery life cycle based on the published literature.

The last chapter's result was extracting and coding the most frequently mentioned factors into five sub-systems and numerous groups based on system thinking. However, redundancy among the elements should be examined before conducting any survey or fieldwork. Not only do overlaps among the criteria cause massive data that are delicate to be analyzed, but they also may easily cause evaluation errors. Furthermore, the authors classified the identified criteria within SNDR projects' life cycle and examined the internal and external interrelationships twice. The classification has been done within of five main stages of SNDR projects' life cycle; "Pre-disaster stage", "post-disaster immediate response stage", "planning and design stage", "procurement, construction, and completion stage", and "continual development stage". As stated in the previous section, the factors have been categorized based on system thinking. After that, the factors were correctly positioned, individual graphs for

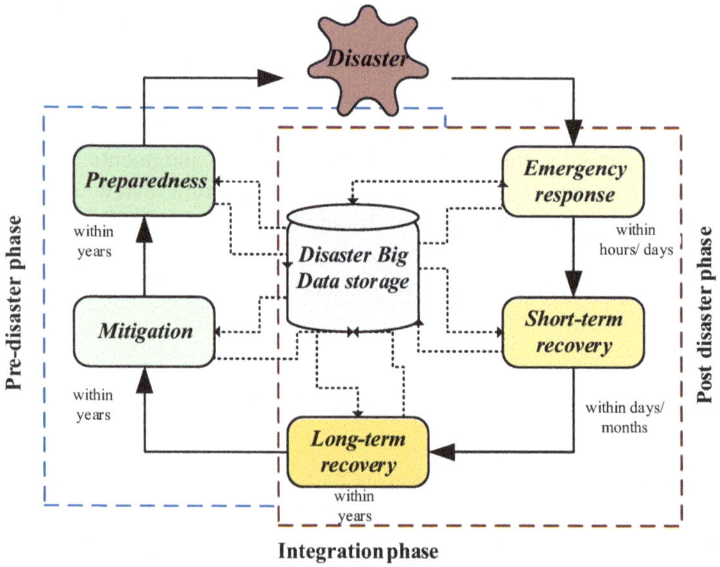

Fig. 4.1 Life cycle of socio-natural disaster recovery projects

Fig. 4.2 Classification of the influential factors based on project life cycle and system thinking

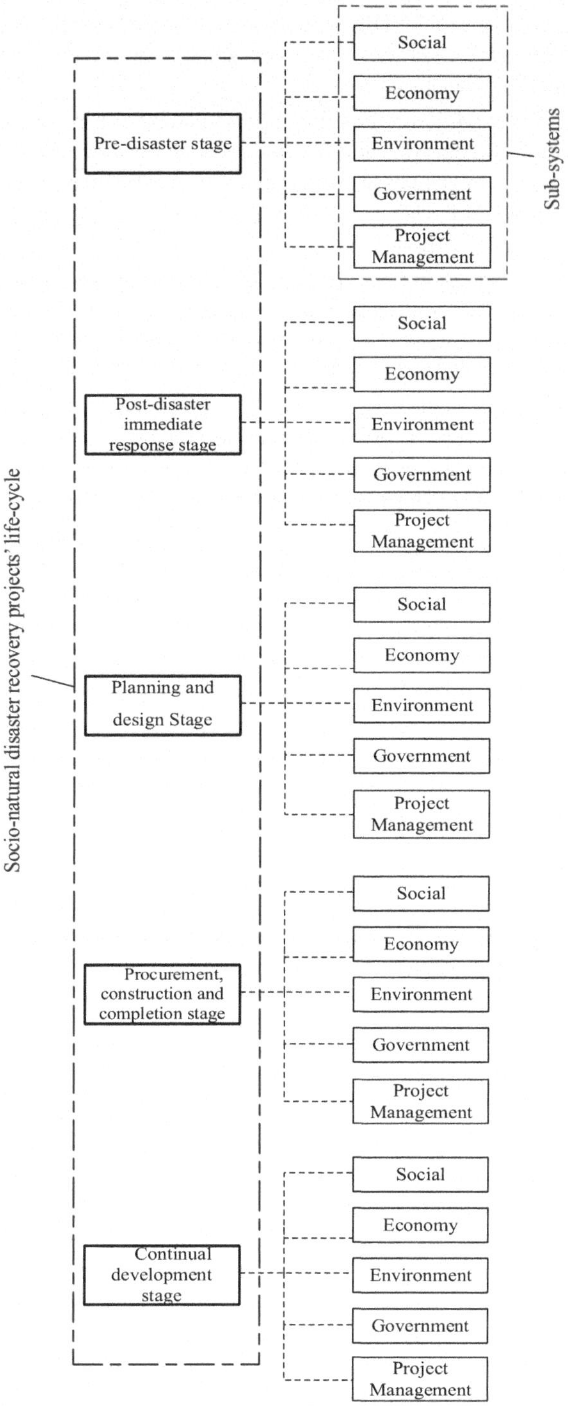

each stage of SNDR projects were drawn, and the interrelationships between the criteria were shown. This process has been done twice to ensure the accuracy of the interrelationships.

Figure 4.2 signifies the applied pattern for the second round of factors' interrelationship analysis and the factors' relationships within each stage. Considering the factors' interrelationship analysis as a refinement process, the effectual elements' validity and relevancy have been examined, and finally, 59 influential factors were extracted. Although this process combined a raft of factors, seven effectual elements have been added or separated from the old ones: "rescue and medical aid", "debris cleaning", "continues disaster handling, budget for reconstruction", "environment protection plan", "use of local construction methods", and "waste management" were out of consideration before factors' interrelationship analysis, and added later (Fig. 4.3). In the end, based on the comprehensive graph, the objectives of the SNDR projects were determined. The nodes that are the influential factors' destinations have been delineated as SNDR projects' objectives. The objectives have formed three groups: short-term, mid-term, and long-term objectives.

"Short-term objectives" that focus on immediate responses after socio-natural disasters. Activities to address community needs and restore damaged facilities have been aimed here. "Mid-term objectives" that mainly focus on the concept of Build Back Better have been divided into delivery, planning, and construction. "Long-term objectives" that represent development after recovery projects assess the fully comprehensive improvement of the community's capability to cope with the next socio-natural disasters. This group's categories are public well-being, the structures' well-being, and procedure well-being. Table 4.1 demonstrates the findings. The main goal of the SNDR project has been identified as the "comprehensive and long-lasting success of the recovery project".

4.2 From Primary Success Factors to Selected Success Factors

Although several researchers have emphasized the literature-based success factors, their reliability within the context of this research should be assessed. Experts' judgment re-examined the PSFs. That is a question that found an answer using the Delphi method. Delphi can be applied to extract information and build consensus upon the primary extracted factors (Hasson et al., 2000). The identified success factors can be validated and refined by taking advantage of unbiased judgments because of anonymous panellists. Additionally, sharing the last round's results with the panellists provides the research with an excellent opportunity to reach a reliable consensus (Diamond et al., 2014). Moreover, due to the panellists' geographical dispersion, the Delphi survey is a practical solution to save time (de Brito et al., 2017).

A list of potential experts, including experts and researchers with experience working on or studying SNDR projects, was made during the content analysis phase.

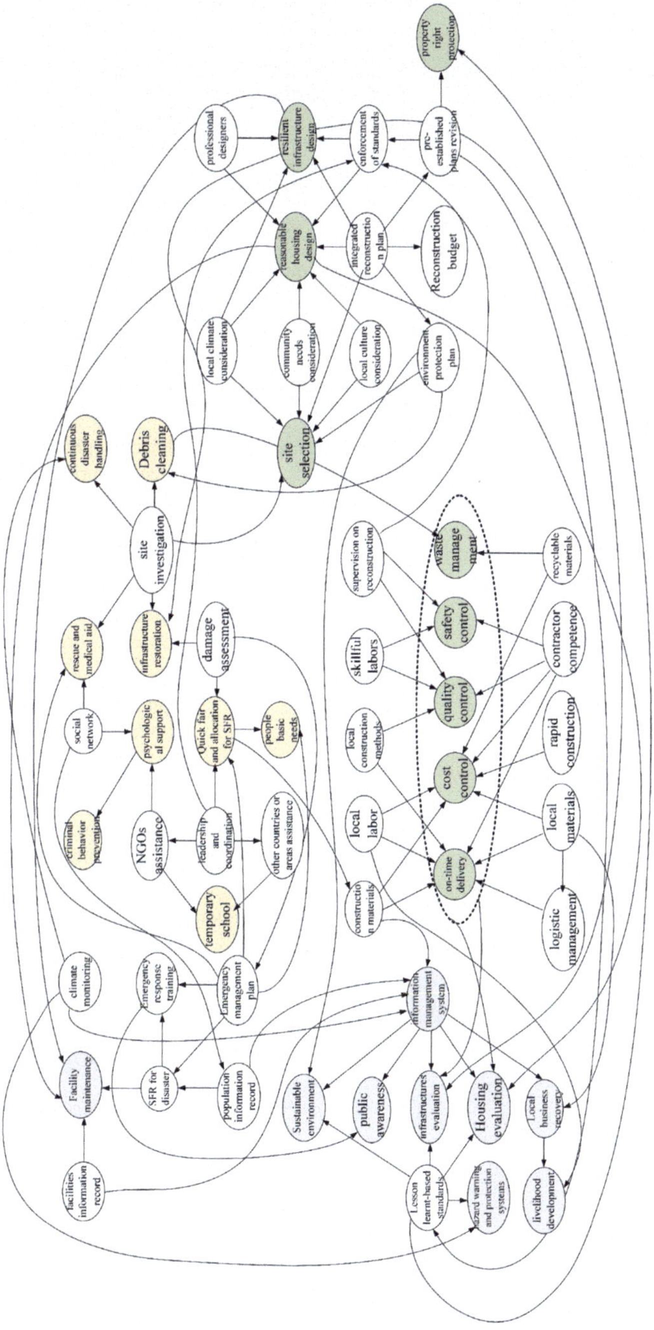

Fig. 4.3 Influential factors within SNDR projects' life cycle and their interrelationships

Table 4.1 SNDR projects' objectives tree (PSFs)

Time-based purpose	First level objectives	Primary success factors	Subsets/influential parameters
Short-term objectives	*Community basic needs*	People's basic needs (food, sanitation, shelter…)	Emergency management plan Quick & fast allocation for SFR
		Temporary school	NGOs assistance Other countries or areas of assistance
		Rescue and medical aid	Social network Site investigation Emergency management plan
Short-term objectives	*Community basic needs*	Psychological support	Social network NGOs assistance
		Criminal behaviour prevention	Psychological support Leadership & coordination
		Quick and fair allocation of SFR	Emergency management plan Damage assessment Leadership & coordination
	Fa cilities & environment	Debris cleaning	Site investigation Environment protection plans
		Continuous disaster handling	Site investigation Climate monitoring
		Immediate infrastructure restoration	Damage assessment Resilient infrastructure design
Mid-term objectives	*Design & planning*	Property right protection	Revision of pre-established plans Updated regulations and standards based on lessons learnt
		Site selection	Integrated reconstruction plan Local climate consideration Community needs consideration Local culture consideration Environment protection plan Site investigation

(continued)

Table 4.1 (continued)

Time-based purpose	First level objectives	Primary success factors	Subsets/influential parameters
		Reasonable housing design	Local climate consideration Community needs consideration Local culture consideration Professional designers Integrated reconstruction plan Enforcement of standards
		Resilient infrastructure design	Integrated reconstruction plan Professional designers Local climate consideration Enforcement of standards
	Construction	Cost control	Use of local labors Availability of construction materials Application of local materials Rapid construction method Contractors' competence Use of recyclable materials
Mid-term objectives	*Construction*	Waste management	Debris cleaning Use of recyclable materials
		Quality control	Use of local construction methods Skillful labors Contractors' competence Supervision on reconstruction
		Safety control	Skillful labors Contractors' competence Supervision on reconstruction
		On-time completion and delivery	Availability of construction materials Use of local labors Logistic management

(continued)

Table 4.1 (continued)

Time-based purpose	First level objectives	Primary success factors	Subsets/influential parameters
			Application of local materials Use of local construction methods Use of recyclable materials
Long-term objectives	*Public well-being*	Improvement of public awareness about natural hazards	Information management system Emergency response training
		Livelihood development	Use of local labors Local business recovery
		Local business recovery	Information management system Application of local materials
		Sustainable environment	Updated regulations and standards Information management system Environment protection plan
	Structures well-being	Facility maintenance	Facilities information record SFR for disasters Reasonable housing design Resilient infrastructures design
		House condition evaluation after reconstruction	Information management system Updated regulations and standards Reasonable housing design Cost control Quality control Waste management Safety control On-time completion and delivery

(continued)

Table 4.1 (continued)

Time-based purpose	First level objectives	Primary success factors	Subsets/influential parameters
		Infrastructure condition evaluation after reconstruction	Resilient infrastructures design Information management system Updated regulations and standards Cost control Quality control Waste management Safety control On-time completion and delivery
	Procedure well-being	Hazard warning and protection systems	Updated regulations and standards based on lessons learnt Climate monitoring

Having a Ph.D. degree in the relevant field AND, publishing at least one academic journal paper about disaster recovery projects, AND having at least three years of related experience in real recovery projects were three prerequisites to be short-listed. The candidates' publication was accessed through academic online search engines. More than one hundred experts have been selected following those criteria. Meanwhile, the number of panellists in Delphi surveys can range from a few to fifty or more, while the most significant population involves several thousand people (Day & Bobeva, 2005). However, at least five experts can reach a consensus of Delphi (Pomponi et al., 2019). The authors first sent an invitation letter to the short-listed potential panellists (briefly describing the research goals, methods, and survey information). If the experts agreed to participate in the survey, they responded through an email. More than one hundred requests have been sent. Finally, ten experts accepted to be panel members, and their qualifications all exceeded the panel selection criteria. Table 4.2 represents the demographic information of the survey participants.

Although the literature suggests the conduction of the four rounds of the Delphi survey, evidence shows that two or three rounds are sufficient (Hasson et al., 2000). For this study, the authors designed two rounds of the Delphi survey. The online questionnaire has been sent to the final short-listed panellists, requiring twenty minutes to be answered. Besides the indicators' rating, the respondents' demographic information was also asked. Moreover, the panellists were required to comment on the definition of the existing factors or add missing factor(s) if they thought there were any. The panellists' comments have been anonymously attached to the results of the following round. The questions consist of fifty-nine PSFs placed in five sub-groups following the logical research framework based on the SNDR projects' life cycle and system thinking. The importance of the factors was assessed through a five-scale Likert scale, as shown in the table 4.3.

Table 4.2 The panellists' demographic information

Panelist age	Panelist study major	Panelist working experience	Country
30–39	Architecture and disaster studies	10–15 years	Iran, Canada
more than 50	Disaster management	More than 16 years	Iran, Australia
30–39	Built environment planning	5–9 years	China, Canada
30–39	Disaster management	5–9 years	Iran
30–39	Disaster management	10–15 years	Iran
30–39	Civil engineer	4 years	USA
more than 50	Earth and Environmental Sciences	More than 16 years	UK
Less than 30	Disaster management	4 years	UK, China
40–49	Architecture and disaster studies	5–9 years	Iran, Japan
more than 50	Housing reconstruction after disasters	10–15 years	Sweden

Table 4.3 Importance rating scale in the Delphi survey

Of little important	Somewhat important	Important	Very important	Extremely important
1	2	3	4	5

As was discussed by some scholars, the consensus analyzing method has to be defined by prior criteria (Diamond et al., 2014). The analysis has been done by calculating the means, maximum, minimum, median, and percentage of agreement. In each round, if the factor's score meets the agreement rule shown in Eq. 4.1, that factor can be put aside and will not appear in the next phase (Nor, 2019). In that formula, X_n represents the experts' score for a specific factor, N is the number of experts, and *freq. of modes* shows the frequency of mode value.

$$Agreementrule = if\left(or\left(\frac{\sum_{i=1}^{n} X_n}{N} \geq 4\right), \left(\frac{freq.ofmode}{N} \geq 0.80\right)\right) \quad (4.1)$$

A new set of questionnaires has been included alongside the last round's report (group's mean, median, maximum, minimum, and percentage of agreement). The first round lasted one month, from 25th November until 20th December 2020. The data that was created throughout the Excel worksheet by the online questionnaire platform (https://www.wjx.cn) was analyzed based on the pre-established criteria described later, and the second-round questionnaire was formed. The second round started immediately and continued until 20th January 2021, while nine panellists attended to rank the refined factors obtained from the last phase.

4.3 Selected Success Factors; Delphi Survey Results

In this section, the authors presented a detailed report of the Delphi results. Every stage of the SNDR project has been assessed separately, and the results of each round are evaluated in the subsequent tables. Not-ranked factors in the first round point to newly added factors in the second phase based on the panellist's comments. To the other extreme, the unexamined factors in the second round, which have gained consensus in the last step, have not had scores in the second round.

4.3.1 Pre-Disaster Stage

Two factors reached the consensus in the first round from the pre-disaster stage and did not appear in the second round. "Emergency management plan" and "emergency response training" have been ranked extremely important by half of the panellists. Interestingly, none of the factors in this stage ranked as "of little importance"; the lowest mean score is 3.3, which belongs to "climate monitoring". The panellists have different opinions on the degree of importance of this factor; however, 40% believe that "climate monitoring" is "important". "Special fund and resource for disasters (SFR)" has shown 40% agreement among the panel who believe this factor is "very important". Based on some panel members' suggestions who said, *"I find "facilities information record" and "facility management" irrelevant"*; those two factors have been combined and formed "vulnerability protection of the built environment". "Facility information record" and "facility maintenance" have been identified as "very important" factors, with 50% and 40% of panellists ranking them, respectively. Similarly, "basic public information", which has been ranked as an "extremely important" factor by 40% of the experts, has been modified to "recognition of residents and business information". Per the panel members' suggestion, "community participation" was added to this stage, and its importance was accessed in the second round.

In the second phase of the survey, "community participation" and "vulnerability protection of the built environment" were agreed upon. The lowest mean again belongs to "climate monitoring", which shows relatively the same voting on different scales. Interestingly, and similar to the first phase's result, no factor is ranked as "of little importance" in this section. The comparison of the results of the two rounds of survey is given in Table 4.4.

4.3.2 Post-Disaster Immediate Response Stage

Table 4.5 provides an overview of the result of post-disaster immediate response. In the first round of the survey, three factors have reached a consensus; "people's

Table 4.4 Results of the Delphi survey, pre-disaster stage

Code	Round	Factors/ Scores	mean	median	mode	max	min	PCT. AGRE %	Consensus
Pre 1	R.1	Basic population information record	3.9	4	5	5	2	40.0	RJCT
	R.2	Recognition of residents and business information	3.7	4	4	5	2	33.3	RJCT
Pre 2	R.1	Special fund and resources (SFR) for disaster	3.7	4	4	5	2	40.0	RJCT
	R.2		3.7	4	4	5	2	44.4	RJCT
Pre 3	R.1	Climate monitoring	3.3	3	3	5	2	40.0	RJCT
	R.2		3.0	3	3	4	2	33.3	RJCT
Pre 4	R.1	Facilities information record	3.7	4	4	5	3	50.0	RJCT
Pre 5	R.1	Facility maintenance	3.7	4	4	5	2	40.0	RJCT
Pre 4&5	R.2	Vulnerability protection of the built environment	4.0	4	4	5	3	55.6	ACCPT
Pre 6	R.1	Emergency management plan	4.4	4.5	5	5	3	50.0	ACCPT
	R.2		–	–	–	–	–	–	–
Pre 7	R.1	Emergency response training	4.1	4.5	5	5	2	50.0	ACCPT
	R.2		–	–	–	–	–	–	–
Pre 8	R.1	Community participation	–	–	–	–	–	–	–
	R.2		4.3	5	5	5	0	55.6	ACCPT

basic needs (food, sanitation, shelter,…)", "rescue and medical aid", and "immediate leadership and coordination". However, because of the comments of two panellists regarding the importance of safe emergency housing after a natural disaster, the first-mentioned factor has been split into two factors: "people's basic needs (food, sanitation, and security)" and "fast provision of safe shelters".

However, these two factors were not accessed in the next round as they had already gained agreement. On the other hand, "temporary school" which shows an average of 2.7 and "assistance from other countries and areas" with a mean of 2.4, were two factors with the lowest average among the panellists. 40% of the panel members

Table 4.5 Results of the Delphi survey, post-disaster immediate response stage

Code	Round	Factors/Scores	mean	median	mode	max	min	P.C.T. AGRE %	Consensus
Post 1	R.1	People's basic needs (food, sanitation, shelter…)	4.4	4.5	5	5	3	50.0	ACCPT
	R.2		–	–	–	–	–	–	–
Post 2	R.1	Temporary school	2.7	3	3	4	1	40.0	RJCT
	R.2		2.7	3	3	4	1	33.3	RJCT
Post 3	R.1	Rescue and medical aid	4.5	4.5	4	5	4	50.0	ACCPT
	R.2		–	–	–	–	–	–	–
Post 4	R.1	Psychological support	3.8	4	4	5	3	60.0	RJCT
	R.2		3.4	4	4	5	2	44.4	RJCT
Post 5	R.1	Criminal behaviour prevention	3.7	4	4	5	1	60.0	RJCT
	R.2		3.8	4	5	5	1	44.4	RJCT
Post 6	R.1	Social network	3.9	4	5	5	2	40.0	RJCT
	R.2	Residents' social networks and trust	3.8	4	4	5	2	44.4	RJCT
Post 7	R.1	Assistance from NGOs	3.3	3.5	4	4	1	50.0	RJCT
	R.2		4.1	4	4	5	4	88.9	ACCPT
Post 8	R.1	Quick and fair allocation of SFR	3.2	3	3	5	2	50.0	RJCT
	R.2		4.0	4	5	5	3	33.3	ACCPT
Post 9	R.1	Debris cleaning	3.5	3	3	5	2	50.0	RJCT
	R.2		3.7	4	4	5	2	33.3	RJCT
Post 10	R.1	Continuous disaster handling	3.5	3.5	4	4	3	50.0	RJCT
	R.2	Consideration of secondary hazards	3.6	3	3	5	2	44.4	RJCT
Post 11	R.1	Immediate leadership and coordination	4	4	5	5	2	40.0	ACCPT
	R.2		–	–	–	–	–	–	–
Post 12	R.1	Immediate infrastructure restoration	3.5	4	4	5	2	50.0	RJCT
	R.2		4.0	4	5	5	2	44.4	ACCPT

(continued)

Table 4.5 (continued)

Code	Round	Factors/Scores	mean	median	mode	max	min	P.C.T. AGRE %	Consensus
Post 13	R.1	Assistance from other countries or areas	2.4	3	3	3	1	60.0	RJCT
	R.2		2.6	2	2	5	1	55.6	RJCT
Post 14	R.1	Damage assessment	3.3	3	3	5	1	50.0	RJCT
	R.2		3.6	4	4	5	2	55.6	RJCT
Post 15	R.1	Site investigation	3.4	3	3	5	1	50.0	RJCT
	R.2		3.8	4	4	5	2	44.4	RJCT
Post 16	R.1	Fast provision of safe shelters	–	–	–	–	–	–	–
	R.2		4.4	4.5	5	5	3	50.0	ACCPT

voted to "temporary schools" as an "important" factor; however, "assistance from other countries and areas" has dominated 60% of the panel members' agreement to consider it "important". Following the panellists' comments, some modifications have been made in the factors' definition. "Continuous disaster handling" has been modified to "consideration of secondary hazards". Similarly, "social network" has been changed to "residents' social network and trust". More than half of the panel members believe that "psychological support" and "criminal behaviour prevention" are "very important"; however, these factors did not meet the consensus criteria.

Three factors met the survey requirements to be counted as agreed influential factors going through the second phase of the survey. Means of the agreed factors, namely "assistance of NGOs", "quick and fair allocation of special funds and resources for disaster", and "immediate infrastructure restoration", are all equal or greater than four, which is one of the criteria for judging consensus. However, 89% of the panellists believe that "assistance of NGOs" is a "very important" influential factor. The lowest means in this section belong to "assistance from other countries or areas" and "temporary schools". The former represents 2.6 as the mean, as more than half of the panellists thought its ranking was "somewhat important". Additionally, 33% of the respondents agreed that the latter one is "important". "Physiological support" with a mean of 3.44, was voted "very important" by 44% of the experts, while only 5% counted it as an "extremely important" criterion.

4.3.3 Planning and Design Stage

In the planning and design stage, the consensus emerged for three factors in the first phase: "consideration of community needs", "budget for reconstruction", and "resilient infrastructure design". As housing is one of the most critical parts of SNDR projects, the authors expected that "reasonable housing design" would gain agreement in the first round; however, the majority of the panellists believe that this factor is "important", while the rest voted to "very important" and "extremely important" equally. Additionally, "environment protection plan" with a mean of 3.5 has been ranked as "important" and "very important" equally. 70% of the panel members ranked "consideration of local culture" as "very important"; however, no agreement has been made upon this factor because one expert has voted on it as "somewhat important". The lowest mean score in this stage is three, which belongs to "designers' professionalism".

According to the second phase of the survey, "integrated recovery plans (simplification of reconstruction procedures)", "site selection", and "reasonable housing design" reached a consensus. Interestingly, there is no factor with a mean less than 3.4, illustrating that almost all the factors in this section are relatively important. Moreover, 66% of the experts believe that "consideration of local culture" is a "very important" factor affecting SNDR projects' success. Similarly, "consideration of local climate" has been voted by 44% of the respondents as an "important" element. "Enforcement of standards", showing a mean of 3.9, close to the research criterion for the agreed factors, has been declared a "very important" item by 55% of the panel members. Table 4.6 walks readers through this stage's results.

4.3.4 Procurement, Construction, and Completion Stage

The results from the first phase of Delphi reveal that one factor in the procurement, construction, and completion stage has gained agreement. "Use of local construction methods" has been voted as an "extremely important" factor by 40% of the panel members and represents the mean of 4.1. Although half of the panel members came to a consensus upon "cost control" and ranked it as "very important", the mean value did not meet the research's criterion. "Rapid construction methods" and "use of recyclable materials" are the factors that showed the lowest mean, 2.6 and 2.7, respectively.

Although the procurement, construction, and completion stage includes several factors, only one factor, "supervision on reconstruction," has been recognized as an agreed influential factor in the second phase of the survey. "Rapid construction methods" represent the lowest mean in this section, while 44% of the respondents have chosen it as an "important" element. The factors related to the project's goals have been expected to reach consensus; however, "cost control," "quality control," "on-time completion and delivery," and "safety control" did not reach the research's set criterion to be selected as final factors. However, "safety control" and "quality control" have presented a high mean degree, 3.9, close to the consensus measurement. The results of this stage are indicated in Table 4.7.

Table 4.6 Results of the Delphi survey, planning and design stage

Code	Round	Factors/Scores	mean	median	mode	max	min	PCT. AGRE %	Consensus
Pla&de 1	R.1	Consideration of local culture	3.8	4	4	5	2	70.0	RJCT
	R.2		3.9	4	4	5	3	66.7	RJCT
Pla&de 2	R.1	Consideration of local climate	3.5	3.5	4	5	2	40.0	RJCT
	R.2		3.6	4	4	5	2	44.4	RJCT
Pla&de 3	R.1	Consideration of community needs	4.3	4	4	5	3	50.0	ACCPT
	R.2		–	–	–	–	–	–	–
Pla&de 4	R.1	Budget for reconstruction	4	4	4	5	3	40.0	ACCPT
	R.2		–	–	–	–	–	–	–
Pla&de 5	R.1	Environment protection plan	3.5	3.5	4	4	3	50.0	RJCT
	R.2		3.4	4	4	4	2	55.6	RJCT
Pla&de 6	R.1	Property right protection	3.8	4	5	5	2	30.0	RJCT
	R.2		3.8	4	4	5	1	44.4	RJCT
Pla&de 7	R.1	Pre-established plans revision	3.7	3.5	3	5	2	40.0	RJCT
	R.2		3.4	3	3	5	2	33.3	RJCT
Pla&de 8	R.1	Integrated recovery plan (simplification of reconstruction procedure)	3.6	4	4	5	2	50.0	RJCT
	R.2		4.1	5	5	5	1	66.7	ACCPT
Pla&de 9	R.1	Designers' professionalism	3	3	4	4	1	40.0	RJCT
	R.2		3.4	4	4	4	1	66.7	RJCT
Pla&de 10	R.1	Site selection	3.7	3.5	3	5	2	40.0	RJCT
	R.2		4.0	4	4	5	1	55.6	ACCPT
Pla&de 11	R.1	Reasonable housing design	3.6	3	3	5	3	60.0	RJCT
	R.2		4.1	4	5	5	2	44.4	ACCPT
Pla&de 12	R.1	Resilient infrastructure design	4.2	4	5	5	3	40.0	ACCPT
	R.2		–	–	–	–	–	–	–
Pla&de 13	R.1	Enforcement of standards	3.7	3.5	5	5	2	40.0	RJCT
	R.2		3.9	4	4	5	2	55.56	RJCT

4.3.5 Continual Development Stage

The last phase of the survey attempted to evaluate the degree of importance of the continual development factors. The experts agreed upon three factors in the first

Table 4.7 Results of the Delphi survey, procurement, construction, and completion stage

Code	Round	Factors/Scores	mean	median	mode	max	min	PCT. AGRE %	Consensus
Pro&con 1	R.1	Use of local labour s	3.9	4	4	5	3	50.0	RJCT
	R.2		3.6	4	4	5	2	33.3	RJCT
Pro&con 2	R.1	Use of local materials	3.8	4	4	5	2	50.0	RJCT
	R.2		3.1	3	4	5	1	33.3	RJCT
Pro&con 3	R.1	Use of local construction methods	4.1	4	5	5	3	40.0	ACCPT
	R.2		–	–	–	–	–	–	–
Pro&con 4	R.1	Cost control	3.5	4	4	5	1	50.0	RJCT
	R.2		3.7	4	4	5	1	44.4	RJCT
Pro&con 5	R.1	Waste management	3.4	3	3	5	2	50.0	RJCT
	R.2		3.1	4	4	4	1	55.6	RJCT
Pro&con 6	R.1	Use of recyclable materials	2.7	3	3	4	2	50.0	RJCT
	R.2		2.8	3	4	4	1	33.3	RJCT
Pro&con 7	R.1	Supervision on reconstruction	3.9	4	5	5	2	40.0	RJCT
	R.2		4.1	5	5	5	1	55.6	ACCPT
Pro&con 8	R.1	Rapid construction method	2.6	2.5	2	4	1	30.0	RJCT
	R.2		2.6	3	3	4	1	44.4	RJCT
Pro&con 9	R.1	Availability of construction materials	3.6	3.5	3	5	2	40.0	RJCT
	R.2		3.8	4	4	5	2	44.4	RJCT
Pro&con 10	R.1	Contractors' competence	3.3	3	3	5	2	40.0	RJCT
	R.2		3.3	3	3	5	2	33.3	RJCT
Pro&con 11	R.1	Skilful labours	3.4	3.5	4	4	2	50.0	RJCT
	R.2		3.6	4	4	5	2	44.4	RJCT
Pro&con 12	R.1	Logistic management	3.4	3.5	4	4	2	50.0	RJCT
	R.2		3.6	3	3	5	2	44.4	RJCT
Pro&con 13	R.1	Safety control	3.7	4	4	5	2	40.0	RJCT
	R.2		3.9	4	4	5	2	55.6	RJCT
Pro&con 14	R.1	Quality control	3.3	3	3	5	2	30.0	RJCT
	R.2		3.9	4	5	5	2	33.3	RJCT
Pro&con 15	R.1	On-time completion and delivery	3	3	4	5	1	30.0	RJCT
	R.2		3.3	4	4	4	2	55.6	RJCT

round; “livelihood development”, “improvement of public awareness about disasters”, and “local business recovery”. However, the authors modified some of the explanations due to the panel’s suggestions. Explanations of “development and recovery of livelihood” and “improvement of public capabilities (awareness) to cope with socio-natural disasters” have been readjusted. “Information management system” showed 60% of agreement and has been voted as “very important”; however, the mean is 3.7, which is lower than the set criterion of consensus. Moreover, “updated regulations and standards based on lessons learned” has reached a mean of 3.9 but still has not gained agreement. Interestingly, no factor has been determined as “of little importance” in this stage.

Although the consensus was reached for three factors in the first phase of the survey, none of the factors gained agreement in the second phase. However, this phase’s means of factors are not lower than 3.6, manifesting their considerable importance. Having a mean of 3.9, “hazard warning and protection system” showed the highest average in this group, while “infrastructures condition evaluation after SNDR” has demonstrated a mean of 3.6, which is the lowest. A visual representation of this stage’s results can be seen in Table 4.8.

4.4 Comparison of Results from Delphi and Critical Literature Review

This section attempts to compare the research findings gathered through two different methodologies. The comparison might lead to identifying the knowledge gaps in the literature, which require further studies. The importance of the factors was evaluated through two methods, Delphi and frequency-based analysis, which were obtained from the literature review. However, to compare those two data sets, the scores should be standardized as one measures the mean value, and the other expresses the frequency rate. By applying fsQCA software, the authors assigned scores ranging from 0–1 to each obtained set. Separated s-curve diagrams for the Delphi results (ranging from 0–5) and frequency of the factors following the literature review (ranging from 1–83) have been generated. The s-curve diagrams presented in Fig. 4.4 and Fig. 4.5 illustrate the data standardization for the Delphi results and literature-based frequencies, respectively. After standardizing the results, the absolute value for the difference between scores has been calculated and presented in Fig. 4.6. The factors not extracted through the literature review (which were recognized during the interrelationship analysis) will not be examined in the comparison (Pla&de 4, Pla&de 5, Pro&con1, Pro&con 3).

The absolute difference values range from 0–0.9. Comparison and categorization of this new data can identify shortcomings in the literature. Recognition of the factors that have reached similar scores from the literature review and the Delphi also provides valuable information emphasizing their importance as some might be highly influential and some have less impact on the SNDR projects. Figure 4.6 demonstrates

Table 4.8 Results of the Delphi survey, continual development stage

Code	Round	Factors/ Scores	mean	median	mode	max	min	PCT. AGRE%	Consensus
ConDev 1	R.1	Livelihood development	4.3	4.5	5	5	3	50.0	ACCPT
	R.2	Development and recovery of livelihood	–	–	–	–	–	–	–
ConDev 2	R.1	Improvement of public awareness about natural hazards	4.2	4	4	5	3	60.0	ACCPT
	R.2	Improvement of public capabilities (awareness) to cope with socio-natural disasters	–	–	–	–	–	–	–
ConDev 3	R.1	Local business recovery	4.3	4	4	5	3	50.0	ACCPT
	R.2		–	–	–	–	–	–	–
ConDev 4	R.1	Sustainable environment	3.7	3.5	3	5	3	50.0	RJCT
	R.2		3.8	4	3	5	3	44.4	RJCT
ConDev 5	R.1	Updated regulations and standards based on lessons learnt	3.9	4	3	5	3	40.0	RJCT
	R.2		3.8	4	4	5	2	66.7	RJCT
ConDev 6	R.1	Information management system	3.7	4	4	5	2	60.0	RJCT
	R.2		3.7	4	3	5	2	33.3	RJCT
ConDev 7	R.1	Hazard warning and protection systems	3.9	4	4	5	3	50.0	RJCT
	R.2		3.9	4	3	5	3	44.4	RJCT
ConDev 8	R.1	House condition evaluation after SNDR	3.6	4	4	5	2	50.0	RJCT
	R.2		3.8	4	4	5	1	55.6	RJCT
ConDev 9	R.1	Infrastructure condition evaluation after SNDR	3.6	3.5	3	5	3	50.0	RJCT
	R.2		3.6	4	3	5	1	33.3	RJCT

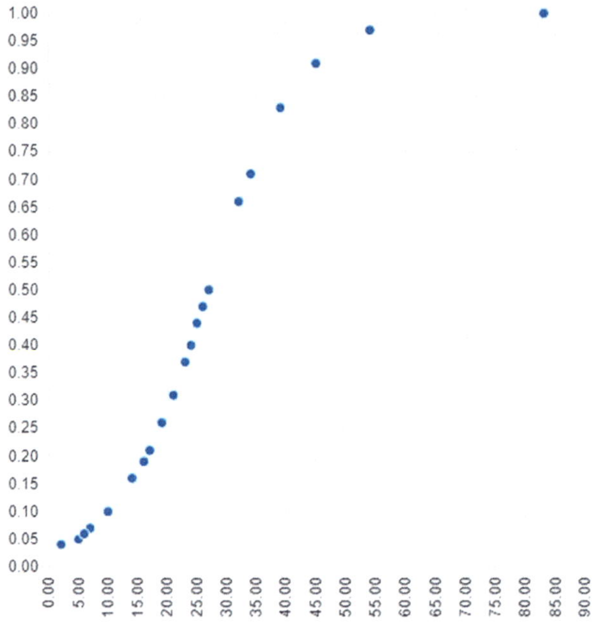

Fig. 4.4 Factor's frequency based on literature review

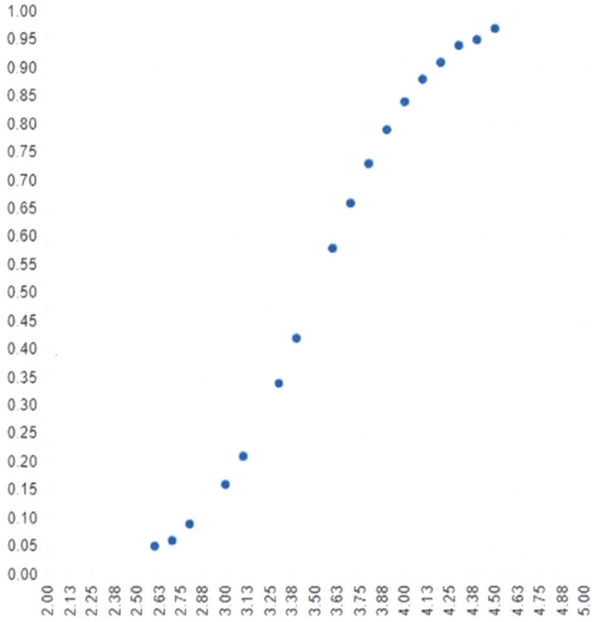

Fig. 4.5 Factor's frequency based on Delphi

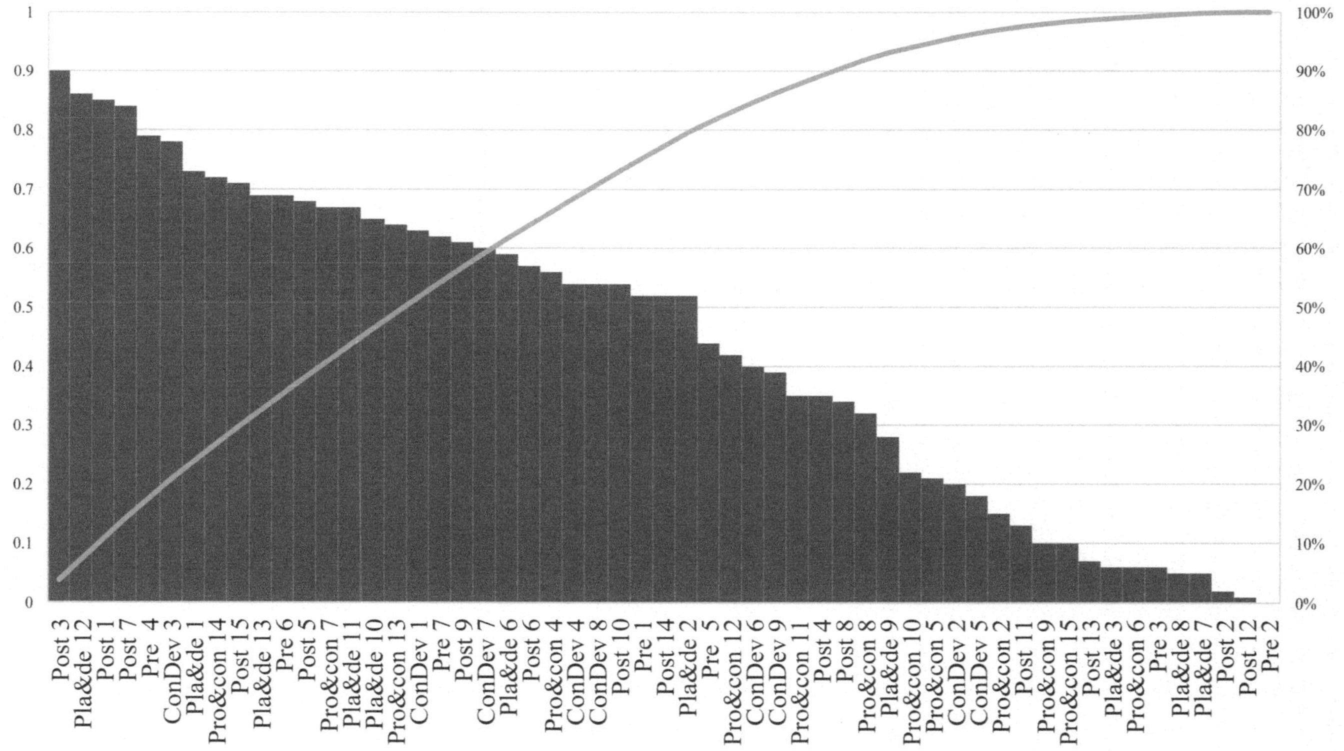

Fig. 4.6 Results' comparison; differences between the normalized values

the top factors resulting from comparing the difference between the critical review and the Delphi. The comparison rules are below: X_i is the standardized value of factor i, obtained from the literature review, and Y_i is that factor's standardized value based on the Delphi result.

$$if \begin{array}{lcl} |Xi - Yi| \leq 0.2 & \rightarrow & The\ most - agreed\ factors \\ |Xi - Yi| \geq 0.7 & \rightarrow & The\ most - non - agreed\ factors \end{array} \tag{4.2}$$

The factors owning the absolute difference values equal to or less than 0.2 have been considered the agreed factors. Interestingly, most of the factors' rankings obtained from the Delphi survey are higher than the literature-based scores except for nine factors. Therefore, the difference values equal to or higher than 0.7 have led to identifying the less-studied topics in the literature. No big absolute difference has been observed in cases that gained higher scores from the literature review, except for "rapid construction method (Pro&con 8)", whose difference value is 0.32. This factor, which has one of the lowest means resulting from Delphi, has frequently appeared in the critical review phase.

4.4.1 Matching Results; the Agreed Factors

Figure 4.7 illustrates the most similar and the most different factors. More than half the factors gained agreement belong to the "post-disaster immediate response stage" and "procurement, construction, and completion stage", which were slightly more emphasized in the literature. "Pre-disaster stage" and "planning and design stage" have the lowest difference scores, but their focus varies as the former has obtained a higher score from the literature, while the latter has been focused more on the survey.

Although "special fund and resources (SFR) for disasters (Pre 2)" has not gained the panel agreement, it has obtained the most similar result as the difference between the two sets of datasets is zero. "Immediate infrastructures' restoration (Post 12)", which reached an agreement following the Delphi survey, also showed a high importance value based on the literature. The third low-difference indicator is "temporary schools (Post 2)", which neither the literature nor the panellists ranked as a significant factor.

On the contrary to "special fund and resources (SFR) for disasters (Pre 2)" having a significant impact on the progress of the socio-natural disaster projects, the establishment of "temporary schools" in the aftermath of a disaster may not greatly influence the recovery. However, further research is needed to examine this finding. "Pre-established plans revision (Pla&de 7)" and "integrated recovery plan (simplification of reconstruction procedure) (Pla&de 8)" gained a fourth place, expressing their role in the management of SNDR projects. However, contrary to pre-established plans revision, which has been more focused on the literature, the integrated recovery plan has gained a higher score based on the Delphi results. "Consideration of community

Codes	Factors gained agreement (Difference less than 0.2)	Difference ranking (Delphi-LR)	Codes	LR gaps (Difference more than 0.70)	Difference ranking
Pre 2	special fund and resources (SFR) for disaster	0	Pre 4	vulnerability protection of the built environment	0.79
Pre 3	climate monitoring	0.06	Post 1	people basic needs (food, sanitation, and security)	0.85
Post 2	temporary school	0.02	Post 3	rescue and medical aid*	0.9
Post 11	Immediate leadership and coordination*	-0.13	Post 7	assistance from NGOs	0.84
Post 12	immediate infrastructure restoration	0.01	Post 15	Site investigation	0.71
Post 13	assistance from other countries or areas	-0.07	Pla&de 1	consideration of local culture	0.73
Pla&de 3	consideration of community needs*	-0.06	Pla&de 12	resilient infrastructure design*	0.86
Pla&de 7	pre-established plans revision	-0.05	Pro&con 14	quality control	0.72
Pla&de 8	integrated recovery plan (simplification of reconstruction procedure)	0.05	ConDev 3	local business recovery*	0.78
Pro&con 2	use of local materials	0.15			
Pro&con 6	use of recyclable materials	-0.06			
Pro&con 9	availability of construction materials	-0.1			
Pro&con 15	on-time completion and delivery	-0.1			
ConDev 5	updated regulations and standards based on lessons learnt	-0.18			

Fig. 4.7 The most similar and the most different factors based on literature analysis and Delphi survey

needs (Pla&de 3)" represents a small difference value; however, it has been counted as an agreed factor following the Delphi result. The significant role of this parameter has widely been focused on as it has earned the highest ranking in the literature review. Although "use of recyclable materials (Pro&con 6)" showed the same difference in consideration of community needs, the literature and the panel did not score highly on the necessity of using recyclable materials.

The panel members have focused more on the role of "climate monitoring (Pre 3)" in recovery projects, as the difference value is 0.06. "Assistance from other countries or areas (Post 13)", "availability of construction materials (Pro&con 9)", and "on-time completion and delivery (Pro&con 15)" manifest the difference ranging from 0.07 to 0.1, all with greater scores resulting from the literature. Among the last-mentioned factors, "assistance from other countries/ areas" has not gained agreement from both the literature and the survey. However, "availability of construction materials" is observed to impact the recovery process greatly, while "on-time completion and delivery" stands in the middle. Although "immediate leadership and coordination (Post 11)" have reached the mean of four from the survey, its importance rate in the literature is higher, causing the difference value of 0.13. However, both results indicate its significance, which should not be neglected. The difference values of "use of local materials (Pro&con 2)" and "updated regulations and standards based on lessons learnt (ConDev 5)" are 0.15 and 0.18, respectively. However, the literature-based score of "updated regulations and standards based on lessons learnt" is higher, contrary to "use of local materials". In addition, "use of local materials" is expected to have a lower effect on the SNDR projects' success than "updated regulations and standards based on lessons learnt".

4.4.2 *Distinctive Results; Needs for Further Studies*

The biggest absolute difference score belongs to "rescue and medical aid (Post 3)". This factor, which had the highest mean score in the Delphi survey, did not gain significant attention in the literature. This finding implies that although not many resources have discussed the role of resuming education after the disaster, the experts' rank stands higher, illustrating the gap in the literature. "Resilient infrastructure's design (Pla&de 12)" shows the second-ranked high difference. Although the panel members have chosen this indicator as one of the most influential factors affecting the recovery process, this indicator has not gained a high importance score from the literature. Even though "consideration of people's basic needs (Post 1)" has been recognized as an obstacle affecting project management, there is still a high demand to study this subject as the literature's score is lower than the survey's result. "Assistance from NGOs (Post 7)", "vulnerability protection of the built environment (Pre 4)", "local business recovery (ConDev 3)", "consideration of local culture (Pla&de 1)", "quality control (Pro&Con 14)", and "site investigation (Post 15)" are the next identified subjects whose role in the management of SNDR projects require further research.

4.5 Chapter Summary

This chapter attempted to define the objectives of successful SNDR projects. The refined influential parameters obtained from the literature review have undergone statistical evaluation to analyze their frequency based on the research's resources. Finally, a comparison of the results of the methodologies was conducted. This chapter's main contributions are as follows:

- This study illustrates a potential objective tree depicting the different possible goals of SNDR projects. Each phase of this time-based objective tree can be used separately as milestones to evaluate the project's progress. Moreover, this study has altered the prevalent unilateral approach toward SNDR projects into a more holistic perspective. The applied systematic logic attempts to answer the heterogeneous needs of stakeholders. Different governments may prioritize the timeline-based categorization of recovery project objectives after socio-natural disasters. Some governments may pay more attention to the short-term goals, while others focus on the long-term ones.

- Based on the Delphi results, while the "continual development stage" has the highest average mean, only 33% of its factors reached a consensus, which illustrates that all the indicators of this group are important. On the other hand, most of the factors in the "pre-disaster stage" have met consensus. Moreover, this stage's mean value indicates the considerable importance of its recovery measurements. Interestingly, the "procurement, construction, and completion stage" shows the lowest average mean. Additionally, few factors from this group reached a consensus, implying that contrary to some research's overemphasis on the reconstruction phase (physical recovery), "procurement, construction, and completion stage" should not be seen as the primary outcome of SNDR projects. The scores of the "post-disaster immediate response stage" and "planning and design stage" are similar. Almost half the factors in each group have reached the set consensus, indicating the importance of these stages' indicators.
- Per the comparison results, the "post-disaster immediate response" stage, owning four factors, shows the highest difference value among the nine identified literature gaps. It could be concluded that the post-disaster immediate response stage, whose accumulative difference value equals 6.74, may have a higher demand to be studied. This stage focuses on the short-term activities in the aftermath of a disaster and includes various responses to answer different groups' needs. On the contrary, the lowest difference belongs to the "pre-disaster stage", whose accumulative difference value is 3.12. Therefore, both the panel members and the literature agreed on the importance of this stage. However, it is necessary to consider the standardized scores alongside the different values.

References

Chen, Y., & Booth, D. C. (2011). The Wenchuan earthquake of 2008 anatomy of a disaster. In *Zhurnal Eksperimental'noi i Teoreticheskoi Fiziki*. Springer. https://doi.org/10.1007/978-3-642-21159-1

Day, J., & Bobeva, M. (2005). A generic toolkit for the successful management of Delphi studies. *Electronic Journal of Business Research Methods, 3*(2), 103–116.

de Brito, M. M., Evers, M., & Höllermann, B. (2017). Prioritization of flood vulnerability, coping capacity and exposure indicators through the Delphi technique: A case study in Taquari-Antas basin, Brazil. *International Journal of Disaster Risk Reduction, 24*(June), 119–128. https://doi.org/10.1016/j.ijdrr.2017.05.027

Diamond, I. R., Grant, R. C., Feldman, B. M., Pencharz, P. B., Ling, S. C., Moore, A. M., & Wales, P. W. (2014). Defining consensus: A systematic review recommends methodologic criteria for reporting of Delphi studies. *Journal of Clinical Epidemiology, 67*(4), 401–409. https://doi.org/10.1016/j.jclinepi.2013.12.002

Dunford, M., & Li, L. (2011). Earthquake reconstruction in Wenchuan: Assessing the state overall plan and addressing the "forgotten phase." *Applied Geography, 31*(3), 998–1009. https://doi.org/10.1016/j.apgeog.2011.01.001

Ganapati, N. E., & Mukherji, A. (2014). Out of Sync: World bank funding for housing recovery, postdisaster planning, and participation. *Natural Hazards Review, 15*(1), 58–73. https://doi.org/10.1061/(asce)nh.1527-6996.0000120

Griffey, A. M. (2010). *Infrastructure disaster management: Insight from small business leaders after a mega disaster* (Issue December). University of Phoenix.

Hasson, F., Keeney, S., & McKenna, H. (2000). Research guidelines for the Delphi survey technique. *Journal of Advanced Nursing, 32*(4), 1008–1015.

Kermanshachi, S., Bergstrand, K., & Rouhanizadeh, B. (2019). *Identifying, weighting and causality modeling of social and economic barriers to rapid infrastructure recovery from natural disasters: A study of hurricanes Harvey, Irma and Maria.*

Nor, M. Z. (2019). Developing a preliminary questionnaire for the faculty development programme needs of medical teachers using Delphi technique. *Journal of Taibah University Medical Sciences, 14*(6), 495–501. https://doi.org/10.1016/j.jtumed.2019.09.007

Pomponi, F., Moghayedi, A., Alshawawreh, L., D'Amico, B., & Windapo, A. (2019). Sustainability of post-disaster and post-conflict sheltering in Africa: What matters? *Sustainable Production and Consumption, 20*, 140–150. https://doi.org/10.1016/j.spc.2019.06.007

Rani, W. N. M. W. M., Nifa, F. A. A., Ismail, M. N., & Khalid, K. N. (2017). Planning for post disaster recovery: Lesson learnt from flood events in Kelantan Malaysia. *AIP Conference Proceedings, 1891*(1), 0201431–1–6. https://doi.org/10.1063/1.5005476

Olshansky, R. (2002). Planning for disasters. *Journal of the American Planning Association, 68*(4), 453.

Supernaw, S. A. (2019). *Impact of project management methodology use in post-disaster recovery and mitigation projects by North Carolina emergency managers* (Issue February). Northcentral University.

Wu, J. Y. (2003). A comparative study of housing reconstruction after two major earthquakes: The 1994 Northridge earthquake in the United States and the 1999 Chi-Chi earthquake in Taiwan.In *ProQuest dissertations and theses* (Vol. 28, Issue August). Texas A&M University.

Xu, J., & Lu, Y. (2018). Towards an earthquake-resilient world: From post-disaster reconstruction to pre-disaster prevention. In *Environmental hazards* (Vol. 17, Issue 4, pp. 269–275). https://doi.org/10.1080/17477891.2018.1500878

Yilmaz, D. G., Meding, J. V., & Erk, G. K. (2013). A theoretical approach to the design of a survey instrument in post-disaster reconstruction: Defining indicators for a human-based study in rural built-environment. *International Journal of Architectural Research, 7*(3), 40–56.

Chapter 5
Identification of Success Dimensions and Critical Success Factors in Socio-Natural Disaster Recovery Projects

Abstract The complex nature of SNDR projects, besides their dynamic nature, requires a dynamic response. The greater the ability to manage the complexity of an SNDR project, the higher the project's chance of success. While traditional project management usually focuses on time, cost, and quality management, SNDR project management should solve the problems arising due to the complexity while addressing the long-term requirements (Chang et al., in JAMA 1(1):65–83, 2010). Therefore, it is necessary to answer the following questions: *What are the success dimensions of SNDR projects?* Moreover, *which factors can be critically influential on the success dimensions?* This chapter attempts to answer the abovementioned questions by clearly defining successful recovery measurements and how influential factors can be used to manage SNDR projects. Triangulation of the results obtained from qualitative and quantitative methods helps the authors to identify causes of success and measurements of successful SNDR projects under the logic of Qualitative Comparative Analysis (QCA). The interactions between the groups of success causes and success measurements have then been assessed. The definition of success is adjustable based on each project's requirements and features. The success dimensions and their influential factors obtained through this chapter are open to further research and can be examined by their application in case studies with distinctive characteristics.

Keywords Critical success factors · Key performance indicators

5.1 List of the Influential Factors in Disaster Recovery Projects

The research has been initiated by exploring academic resources found in academic search engines such as Science Direct, ProQuest, and Springer. While publications after 2003 written in English focused on "disaster recovery" AND "natural disasters" were the first keywords used to gain the first results. Later, post-natural disaster

H. Bahmani et al., *Socio-Natural Disaster Recovery Projects*, Resilient and Friendly Living Environments, https://doi.org/10.1007/978-981-95-0113-7_5

recovery and post-disaster reconstruction refined the results, and the authors selectively chose the most relevant publications. Subsequently, the potential indicators were extracted through a detailed review of the selected literature and codified based on system thinking in NVivo2020. Afterwards, a continuous life cycle for the SNDR project offered based on the published literature was applied to evaluate the factors' interrelationships. Finally, 59 influential factors have been identified and presented in Table 5.1.

Afterwards, the number of the received/ sent links among the influential factors based on factors' interrelationship analysis and the accumulative frequency of the final effectual elements based on the literature has been shown in Fig. 5.1. The frequency of some factors based on literature has been combined, as they have not been extracted from the literature. These parameters were later added during the factors' interrelationship analysis.

Based on the critical literature review, "consideration of community needs", "immediate leadership and coordination", "updated regulations and standards based on lessons learnt", "integrated recovery plan (simplification of recovery procedure)", "availability of construction materials", "immediate infrastructures restoration", "improvement of public awareness about natural hazards", and "special fund and resources for disasters" are the highly-ranked influential factors. "Planning and design stage", which has shown the highest accumulative ranking of its elements, is the most crowded stage; however, the "pre-disaster stage" owns the lowest accumulative frequency of the influential factors.

The Delphi technique has been chosen to check the reliability of the extracted factors in Table 5.1. Additionally, the panellists have been asked to modify the factors, and their anonymous comments have been attached to the second round of the questionnaire alongside the first round's results. The indicators that met the criteria for consensus have been put aside and not evaluated later. Finally, the panel agreed upon twenty-two factors, as shown in the Table 5.2.

5.2 Structuring Success Dimensions and Their Influential Factors

This section is devoted to presenting the analysis of the results obtained earlier in chapter 4. The highly-ranked factors from each method (critical review, factors' interrelationship analysis, and Delphi) were identified to be compared by triangulation. The authors established three criteria for choosing the top factors in each phase. Twenty-five top influential factors have been listed based on the literature's frequency analysis (Fig. 5.1), and the parameters with equal or more than four links formed the second part (Fig. 5.1). Finally, the factors that have reached consensus in the Delphi survey have been selected as the third part of this comparison (Table 5.1). Figure 5.2 shows the scenario explained above.

Table 5.1 Influential factors resulted from literature review and factors' interrelationship analysis

Life cycle stage	Code	Factors	Code	Factors
Pre-disaster stage	PRE1	Recognition of residents and business information	PRE6	Emergency management plan
	PRE2	Special fund and resources (SFR) for disaster	PRE7	Emergency response training
	PRE3	Climate monitoring	PRE8	Community participation
	PRE 4&5	Vulnerability protection of the built environment		
Post-disaster immediate response stage	Post 1	People's basic needs (food, sanitation, and security)	Post 9	Debris cleaning
	Post 2	Temporary school	Post 10	Consideration of secondary hazards
	Post 3	Rescue and medical aid	Post 11	Immediate leadership and coordination
	Post 4	Psychological support	Post 12	Immediate infrastructure restoration
	Post 5	Criminal behaviour prevention	Post 13	Assistance from other countries or areas
	Post 6	Residents' social networks and trust	Post 14	Damage assessment
	Post 7	Assistance from NGOs	Post 15	Site investigation
	Post 8	Quick and fair allocation of SFR	Post 16	Fast provision of safe shelters
Planning and design Stage	Pla&de 1	Consideration of local culture	Pla&de 8	Integrated recovery plan (simplification of reconstruction procedure)
	Pla&de 2	Consideration of local climate	Pla&de 9	Designers' professionalism
	Pla&de 3	Consideration of community needs	Pla&de 10	Site selection
	Pla&de 4	Budget for reconstruction	Pla&de 11	Reasonable housing design
	Pla&de 5	Environment protection plan	Pla&de 12	Resilient infrastructure design*
	Pla&de 6	Property right protection	Pla&de 13	Enforcement of standards
	Pla&de 7	Pre-established plans revision		
Procurement, construction, and completion stage	Pro&con 1	Use of local labors	Pro&con 9	Availability of construction materials

(continued)

Table 5.1 (continued)

Life cycle stage	Code	Factors	Code	Factors
	Pro&con 2	Use of local materials	Pro&con 10	Contractors' competence
	Pro&con 3	Use of local construction methods	Pro&con 11	Skilful labours
	Pro&con 4	Cost control	Pro&con 12	Logistic management
	Pro&con 5	Waste management	Pro&con 13	Safety control
	Pro&con 6	Use of recyclable materials	Pro&con 14	Quality control
	Pro&con 7	Supervision on reconstruction	Pro&con 15	On-time completion and delivery
	Pro&con 8	Rapid construction method		
Continual development stage	ConDev 1	Development and recovery of livelihood	ConDev 6	Information management system
	ConDev 2	Improvement of public capabilities (awareness) to cope with natural hazards	ConDev 7	Hazard warning and protection systems
	ConDev 3	Local business recovery	ConDev 8	House condition evaluation after SNDR
	ConDev 4	Sustainable environment	ConDev 9	Infrastructure condition evaluation after SNDR
	ConDev 5	Updated regulations and standards based on lessons learnt		

Seven factors have appeared in all the methods utilized in triangulation. "Vulnerability protection of the built environment", "emergency management plan", "quick and fair allocation of special funds and resources", "immediate leadership and coordination", "integrated recovery plan (simplification of reconstruction procedure)", "site selection", and "reasonable housing design" have gained consensus from all the applied triangulation methods (named as 3/3 in Fig. 5.2). However, the rest of the influential factors can be grouped in seven clusters; "simultaneously ranked by literature review and factors' interrelationship analysis", "simultaneously ranked by literature review and Delphi", "simultaneously ranked by Delphi and factors' interrelationship analysis", "ranked by literature review", "ranked by factors' interrelationship analysis", "ranked by Delphi", and "non-ranked factors". The criteria voted in two sections simultaneously (named 2/3 in Fig. 5.2) show relatively higher importance than those selected by only one method (named 1/3 in Fig. 5.2). Similarly, the non-ranked factors (named 0/3 in Fig. 5.2) represent the lowest degree of importance among all the identified effectual factors. This analysis is the basis

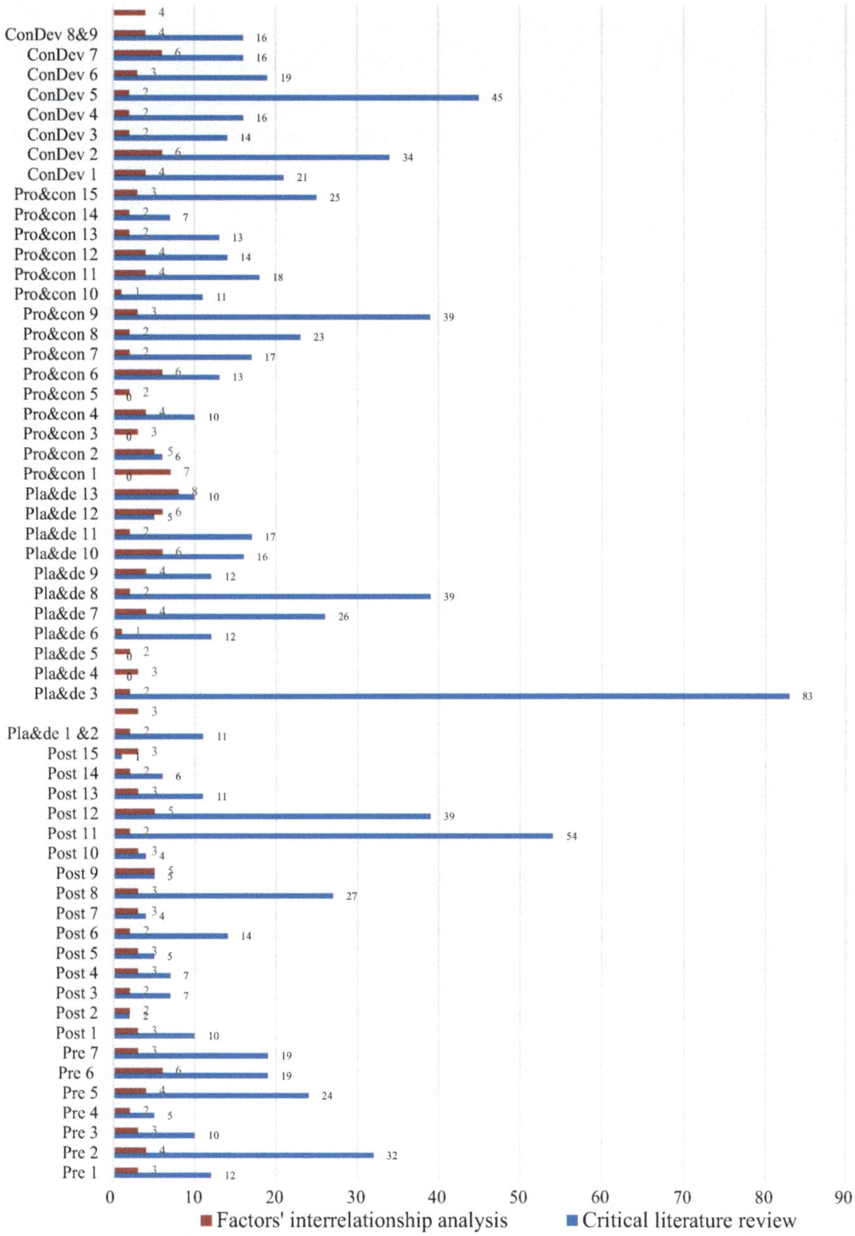

Fig. 5.1 Final influential factors' frequency based on critical review literature and factors' interrelationship analysis

Table 5.2 Agreed factors resulted from the Delphi survey

First-round agreed factors				Second-round agreed factors			
Code	Mean	Mode	Agreement %	Code	Mean	Mode	Agreement %
Pre 6	4.4	5	50.00	*Pre 4&5*	4	4	55.56
Pre 7	4.1	5	50.00	*Pre 8*	4.33	5	55.56
Post 3	4.5	4	50.00	*Post 1*	4.4	5	50.00
Post 11	4	5	40.00	*Post 7*	4.11	4	88.89
Post 16	4.37	5	50.00	*Post 8*	4	5	33.33
Pla&de 3	4.3	4	50.00	*Post 12*	4	5	44.44
Pla&de 4	4	4	40.00	*Pla&de 8*	4.11	5	66.67
Pla&de 12	4.2	5	40.00	*Pla&de 10*	4	4	55.56
Pro&con 3	4.1	5	40.00	*Pla&de 11*	4.11	5	44.44
ConDev 1	4.3	5	50.00	*Pro&con 7*	4.11	5	55.56
ConDev 2	4.2	4	60.00				
ConDev 3	4.3	5	50.00				

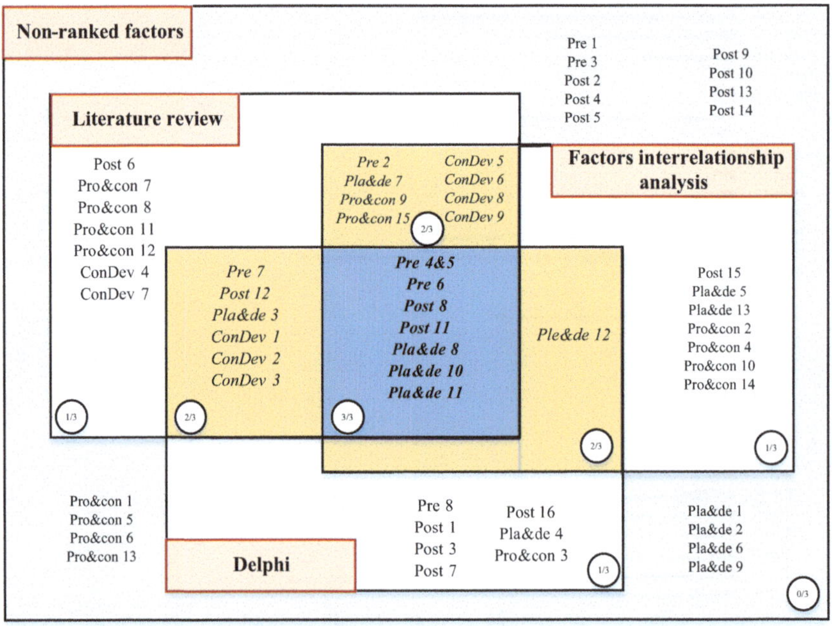

Fig. 5.2 Results' clusters generated by triangulation methodology (3/3: Full agreement, 2/3: 2/3 agreement, 1/3: 1/3 agreement, 0/3: Non-agreement)

for creating measurable success factors and defining success dimensions in SNDR projects.

Three necessary steps should be taken to apply QCA and structure success dimensions and their influential factors. First, outcome(s) should be determined. The outcome is the procedure's objective or result (desirable or undesirable). Second, the characteristics that affect the outcomes should be defined. These characteristics, which are going to be assessed through cases, are called conditions. Conditions should be selected based on the existing theory, and the researchers might try to choose the most relevant ones to avoid adding more complexity to the model (Jordan, 2012). The term condition is somewhat similar to "variable" used in quantitative methods (Devers et al., 2013). Finally, cases should be carefully selected to include a variety of outcomes within cases. Random case selection is expected in large-N research, which may bring heterogeneous cases into the sample and should be avoided in QCA (Rihoux, 2006). Subsequently, the required data for the conditions and outcome(s) were collected, and through the process of calibration and sub-set analysis, combinations of conditions leading to the outcome(s) were analyzed. This procedure will be discussed in chapter 7).

In this study, the vital parameters affecting SNDR projects' success regarded as CSFs are called "conditions". Meanwhile, a successful SNDR project's ultimate goal(s), which are Key Performance Indicators (KPIs), can be defined as "outcome(s)". The triangulation of the results from the last section smoothed the path of recognizing these parameters. According to the triangulation comparison results, this division groups the outcomes and conditions of SNDR projects. A time-based life cycle, which has been proposed before, has been applied to form the conditions. The conditions have been grouped into short-term, mid-term, and long-term periods. As most of the current literature pointed out the importance of alleviating vulnerabilities and systems' quick ability to bounce back to normalcy after disasters (Edgington, 2011; Han, 2014; He, 2019), the definition of sustainability and resiliency alongside system thinking generated the outcomes' table.

Table 5.3 and Table 5.4 separately lists the outcomes and conditions. As shown in those tables, each factor is accompanied by multiple indices to measure that factor. The criteria ranked by none of the triangulation methods formed indices of outcomes or conditions, and the rest of the factors may be used as main conditions/ outcomes. However, some factors that have gained agreement in at least one methodology have been grouped as indices of the foremost parameters to simplify the model. According to the literature, some extra indices were also used (Cutter et al., 2010). The authors later collected data to assess the criteria through surveys and secondary documents within case studies.

Table 5.3 Table of outcomes (KPIs)

Systematic outcomes	No	Outcomes	Indices
Social recovery	A_1	Resilient Society	Social connections
			Phycological support (suicide, mental disorders report
			Safety (criminal behavior)
			Life satisfaction
			Satisfaction of recovery process
			Population growth
			Equity among population
Environment recovery	A_2	Sustainable and resilient built environment	Safety of construction methods and technologies
			Homeless people
			Unoccupied houses
			Population per capita in houses
			Hygiene water accessibility
			Electricity accessibility
			Roads improvement
			Public transportation
			Number of schools, hospitals
			Evacuation facilities
	A_3	Sustainable natural environment	Environmental pollution
			Environmental-friendly construction
Economy recovery	A_4	Resilient economy	Number of new businesses initiation or restoration of old ones
			GDP growth rate
			Employment growth rate
			Household income growth rate
Policy recovery	A_5	Sustainable policy development	Simplified procedures
			Regular revision of plans
			Simultaneous consideration of pre-and post-disaster plans

5.3 Key Performance Indicators (Outcomes) Based on System Thinking

The holistic perspective gained by the system thinking approach has been combined with the definition of sustainability and resilience to set the SNDR projects' goals. A model for assessing community resiliency may consist of social, economic, institutional, infrastructure, and community capacities (Cutter et al., 2010). However, in

Table 5.4 Table of conditions (CSFs)

Time-based framework	No	CSFs (conditions)	Indices
Short-term	a_1	Sufficient answer to people's basic needs*	Food availability
			Sanitation level
			Starting time for rescue and site investigation
			Beneficiary selection
			Sufficient and on-time budget allocation
			Immediate leadership & coordination
			NGOs assistance
			Other areas assistance
	a_2	Availability of shelters and schools	Shelter availability
			School availability
	a_3	Immediate restoration of environment and infrastructures	Damage assessment
			Debris cleaning
			Climate monitoring
Mid-term	a_4	Reasonable housing design*	Consideration of local culture
			Consideration of local climate
			Consideration of community needs
			Use of local construction methods
	a_4	Reasonable housing design*	Professionalism of designers
			Site selection (distance to facilities and safety)
			Consideration of standards
	a_5	Resilient infrastructure design*	Consideration of local climate
			Professionalism of designer
			Consideration of standards
	a_6	Project management***	Be on the planned budget
			Immediate leadership and coordination
			On-time completion and delivery
			Quality control
			Supervision on reconstruction
			Labour force's price control
			Contractors' competence

(continued)

Table 5.4 (continued)

Time-based framework	No	CSFs (conditions)	Indices
Mid-term	a_6	Project management***	Availability of skilful labour force
			Logistic management
			Rapid construction methods
			Availability of materials
Long-term	a_7	Improvement of public capabilities	Emergency response training
			Safe construction training
			Livelihood development
	a_8	Community engagement level	Decision-making
			Planning
			Construction
	a_9	Local economy improvement	Recovery/development plans for local business
			Use of local materials and labour force
	a_{10}	Integrated recovery policies	Pre-established plans
			Emergency management plans
			Updated regulations and standards per lessons learned
			Enforcement of standards
			Property right legislation
	a_{11}	Vulnerability protection of the built environment	Regular maintenance
			Information management system
			Hazard warning and protection systems
			Consideration of secondary hazards

* Sufficient answers to people's basic needs cover emergency needs and fund and assistance management
** Reasonable housing design includes indicators covering community and technical design
*** Project management consists of planning and construction management

this research, system thinking logic divided the outcomes into four stages: social recovery, environment recovery, economic recovery, and policy recovery. Subsequently, each group consists of primary outcomes based on the definition of resiliency or sustainability, or both.

SNDR projects should be able to reduce the possibility of upcoming risks and susceptibilities. Developing capabilities throughout resiliency improvement should also be planned (Xu et al., 2014). While sustainability evaluates results, resiliency can assess the process (Shao, 2017). Another difference between resiliency and sustainability is lifetime measures. While sustainability looks forward to reducing vulnerabilities in the long-term future, resiliency focuses on the systems' capability to return to normalcy in a short period (Rose, 2011). However, instead of oversimplifying the concept of resiliency to come back to the pre-disaster level, which may generate the previous vulnerabilities, increasing communities' disaster resiliency has been suggested (Lu & Xu, 2015). Recent studies have usually seen resiliency in the following phases: preparedness, absorption, recovery, and adaption (Bostick et al., 2018). Robustness, rapidity, and the system's ability to return to customary conditions have also been seen as resiliency functions (Barabadi & Ayele, 2018). The following paragraphs provide a brief explanation of each outcome in Table 5.3.

(1) Resilient Society

Different indicators can measure this outcome, referring to a community's ability to return to normal situations (Xu & Lu, 2018). The multiplication of the different factors can be used to evaluate social resilience, such as annual population growth, annual growth of GDP, and Gini index of equality (Platt, 2018). Another study found sex ratio, per capita GDP, percentage of ethnic minorities, and medical facilities as the most influential characteristics of social resilience (Li et al., 2016). However, in this study, some different indicators have been set. One of the objectives of SNDR projects is restructuring community ties and social networks (Sadiqi et al., 2017a), which can be counted as an indicator of a resilient society. Although psychological support and criminal behaviour protection are non-ranked factors based on triangulation, some scholars count them as essential recovery steps (Han, 2014). Residents' satisfaction with life and the recovery process, population growth, and equity among the population (especially in participation and aid allocation) are the other indicators assisting the analysis of this outcome.

(2) Sustainable and Resilient Built Environment

This outcome can be measured by considering indicators from two possible groups: housing and infrastructure. Sustainable reconstruction has often been considered a systematic development methodology focusing on long-term impacts. Houses, where people spend most of their time, should be sustainably designed to stand against socio-natural disasters (Roosli & O'Keefe, 2013). Records of homeless people after the completion time of the recovery project and unoccupied newly-built houses alongside people's accessibility to water and electricity in houses have set this outcome's indicators. On the other hand, the reconstruction of structures, especially infrastructures, applies modern technologies to enhance infrastructures' resilience and withstand a defined level of future disasters (Hu et al., 2018; Rani et al., 2017). The improvement or decline in public transportation, roads, and public social well-being service centres have been structured to evaluate the group of infrastructures.

(3) Sustainable Natural Environment

One aspect of sustainability is the perception of its impacts on the natural environment (Tucker et al., 2014). As a result, the degree of environmental pollution after SNDR projects has been established as one indicator. In addition, strategies to mitigate the adverse effects of reconstruction on the natural environment (Silva, 2010), generally called environmentally friendly construction, measure this outcome's score.

(4) Resilient Economy Okamoto

A resistant economy, which can tolerate fluctuations and adverse disaster effects, can be assessed through several indicators. Numerous studies utilized GDP as an indicator measuring society's resiliency (Li et al., 2018; Platt, 2018); however, to collect accurate data, the authors established other indicators. The growth rate of employment, household income, and increase/ reduction in the number of new business initiations or restoration of the old business has set the indicators of this KPI.

(5) Sustainable Policy Development

(Okamoto & Ishikawa, 2013) proposed that sustainability is divided into three groups: environmental, cultural, and social. Taking the concept of "social sustainability", which expresses the restoration of decision-making, this research has modified this group's name to "sustainable policy development" to cover more comprehensive topics. Sustainable policy can be appraised through planning strategies. Generally, these strategies should be followed in pre-disaster and post-disaster recovery plans, which might be established and utilized in a cyclic. Furthermore, the authors have chosen simplification of the procedures, especially in permission obtainment (Di Gregorio & Soares, 2017; Du et al., 2016) and consideration of planning revision (Griffey, 2010), as the indices of this outcome.

5.4 Critical Success Factors (Conditions) Based on Continual Life-Cycle

The influential factors, also named CSFs, are assumed to affect recovery projects' success after socio-natural disasters. As shown in Table 5.4, time-based objectives are depicted based on the continuous life cycle of SNDR projects. QCA will generate a series of tables whose data amount is directly related to the number of conditions and outcomes. Therefore, keeping the number of conditions and outcomes low is a (Berg-Schlosser & Meur, 2012). Each stage of this time-based table includes criteria and indices. The indices ease the measurement of the factors. The following paragraphs briefly explain the conditions and indices in each group, referring to the time-based framework.

(1) Short-Term CSFs

The community's urgent needs and the fast revitalization of critical infrastructures are the conditions in this group. Based on the Delphi results, people's basic needs and rescue and medical aids have reached the panel consensus. Referring to the literature, people's basic needs must be answered to restore the lifeline service (Lizarralde,

2004). Additionally, the importance of emergency aid and recovery funding, which is a potential subject affecting SNDR projects' success, has been focused on in some studies (Zhang & Peacock, 2010). On the other hand, the restoration of the infrastructure that suffered and its effect on social recovery has gained attention in the literature (Nejat & Ghosh, 2016). In addition, following the Delphi results, half of the panel ranked it as a crucial influential factor. Therefore, damage assessment, debris cleaning, and climate monitoring have formed the indices of this condition.

(2) Mid-Term CSFs

Influential factors related to housing, infrastructures, and project management have been grouped and structured in the mid-term division. This group, which implies the importance of design and construction phases, has been investigated by several researchers. Both literature and the triangulation results have mentioned consideration of local culture, climate, and residents' needs, alongside the competence of designers and safe and acceptable site selection (Silva, 2010; Tauber, 2015). Similarly, resilient infrastructure design, which gained acceptance from two applied methods in triangulation, consists of technical indicators. Finally, project management, structured based on the relevant recognized indicators, consists of planning control and construction control. Time, cost, quality management, and immediate leadership and supervision of reconstruction are the indicators of planning control. On the other hand, the construction control group delves into addressing on-site topics such as material and methods, labour force, and logistics management.

(3) Long-Term CSFs

The effects of improved social well-being, economy, environment, decision-making process and legislation, and buildings have been evaluated by this condition. People, as the main consumers of the recovery outcomes, should be informed and trained. The necessity of considering community involvement, especially in developing countries, has been the other critical condition affecting the recovery outcomes (Sadiqi et al., 2017a; Xu et al., 2014). On the other hand, as having a stable occupation is the main element of each family's livelihood recovery, providing carriers for residents and hiring them can facilitate local economy recovery (Olshansky, 2006). Integrated recovery policies consist of structuring and revising the pre-established emergency and recovery plans, establishing property rights legislation, and enforcing the set standards. Vulnerability protection of the built environment, agreed by all the triangulation methods, consists of applying a hazard warning system, considering secondary hazards, using an information management system, and conducting regular maintenance.

5.5 Interaction Model of CSFs and KPIs

Considering conditions as CSFs in SNDR projects, the outcome(s) results from the combination(s) of conditions. Although the research has already recognized the KPIs (outcomes) and CSFs (conditions), not all the conditions may directly affect the outcomes (Some may not have theoretical links) (Jordan, 2012). Furthermore, it is crucial to formulate the interactions between CSFs and KPIs before going further to avoid adding more complexity to the interaction models. The data collected through context analysis is the main tool used to determine the theoretical links between conditions and outcomes. The background of the theoretical links is given in the following paragraphs.

According to the definition of successful recovery projects after socio-natural disasters, the success of SNDR projects can be measured through five aspects. Furthermore, the interactions between CSFs and KPIs can be depicted as shown in Fig. 5.3. As the identified conditions CSFs may influence each KPI, five primary models between KPIs and CSFs can be considered. Each model has focused on assessing the project outcome based on the specific assigned KPI. Finally, those five models' interactions can comprehensively assess the long-lasting success of the SNDR projects. The presented model of the interactions between the identified KPIs and CSFs depicts the general relationship in SNDR projects regardless of the project's content. The recognized KPIs and CSFs can be specified for different types of socio-natural disasters and variable scopes.

(1) Outcome A_1

As the measurement of a resilient society is based on people's social ties, satisfaction, and population growth, the authors considered the conditions that can directly impact this outcome. Furthermore, the conditions related to housing, infrastructures, people empowerment, and policy-making significantly impact a resilient society. Although most conditions can influence this outcome, no theoretical link has been found between this outcome and two conditions: project management and local economic improvement.

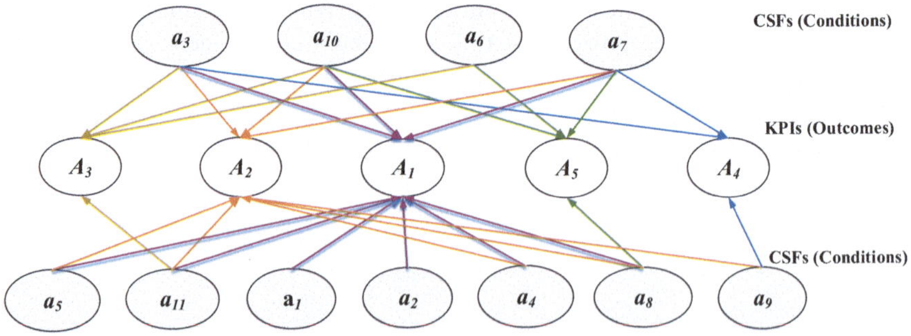

Fig. 5.3 Interactions between outcomes and conditions

(2) Outcome A_2

Indicators that appraise sustainable and resilient built environments can be divided into housing and infrastructure conditions. Therefore, theoretical links have been found between the conditions pointing to these groups. Furthermore, except for three conditions, a_1, a_2, and a_6, related to emergency response and economic recovery, the rest of the conditions have been assumed to affect this outcome directly.

(3) Outcome A_3

The sustainable natural environment has two indicators, environmental-friendly construction and environmental pollution, and it has relationships with the effects of reconstruction and procurement, damage assessment and debris cleaning, buildings' maintenance, and recovery plans. However, the existing literature shows that conditions pertinent to community needs and empowerment, facility design, and economic recovery do not contribute to forming a sustainable natural environment. As a result, seven conditions have been recognized as having a straight impact on this outcome.

(4) Outcome A_4

Resilient economy and business restoration have been directly linked to the fast restoration of the infrastructures, especially the transportation system (Griffey, 2010). Moreover, livelihood development and business recovery plans undeniably affect this outcome. However, a wide variety of conditions do not affect the resilient economy. In general, those conditions are discussed in terms of social and technical aspects.

(5) Outcome A_5

People and planners are two stakeholders that can significantly impact sustainable policy development. In addition, the other aspect influencing this outcome is the role of a trade-off between the project goals, time, cost, quality, planning, and management. However, some conditions may not affect this outcome. The conditions debating technical, economic, and environmental topics can be excluded from the list of influential conditions.

5.6 Chapter Summary

This chapter's key idea is to use triangulation methodology and QCA to identify success dimensions and their influential factors for SNDR projects. The following paragraphs express the study's significance:

- As the research attempts to consider SNDR projects as a complex system that includes various fields, the recognized outcomes in this framework can be viewed separately or together to examine the project's degree of success. The outcomes aim to examine different stakeholders' perspectives on the recovery process. Some of the outcomes point to public opinion, while others might need decision-makers participation or using official reports to be assessed. The KPIs and CSFs and their interaction model can be utilized to evaluate any type of SNDR project's success. However, it is essential to modify the components based on the context of cases as some of the factors might be missing or unmeasurable depending on the situation of that project.

- Furthermore, the findings and interaction model can be examined via multiple case studies within a specific context to show the necessity and sufficiency of factors (or a combination of them) to reach the outcomes (Chapter 6). Moreover, outcomes can establish checklists to identify the project process's vital parameters. The checklists, which are management facilitators, can measure the project management effectiveness.

The current findings suggest several avenues for future research. Socio-natural disaster recovery projects are naturally complex, and a combination of elements might generate different results. Furthermore, future research should focus on examining the various cases to define the most important influential factors. Applying other methods to analyze the instances within different contexts can optimize the results and propose project management guiltiness for SNDR projects within different regions.

References

Barabadi, A., & Ayele, Y. Z. (2018). Post-disaster infrastructure recovery: Prediction of recovery rate using historical data. *Reliability Engineering and System Safety, 169*, 209–223 (August 2017). https://doi.org/10.1016/j.ress.2017.08.018

Berg-Schlosser, D., & Meur, G. De. (2012). Comparative Research Design: Case and Variable Selection. In B. Rihoux & C. C. Ragin (Eds.), *Configurational Comparative Methods: Qualitative Comparative Analysis (QCA) and Related Techniques* (pp. 19–32). SAGE Publications, Inc. https://doi.org/10.4135/9781452226569.n2

Bostick, T. P., Connelly, E. B., Lambert, J. H., & Linkov, I. (2018). Resilience science, policy and investment for civil infrastructure. *Reliability Engineering and System Safety, 175*, 19–23. https://doi.org/10.1016/j.ress.2018.02.025

Chang, Y., Wilkinson, S., Seville, E., & Potangaroa, R. (2010). Resourcing for a resilient post-disaster reconstruction environment. *International Journal of Disaster Resilience in the Built Environment, 1*(1), 65–83. https://doi.org/10.1108/17595901011026481

Cutter, S. L., Burton, C. G., & Emrich, C. T. (2010). Disaster resilience indicators for benchmarking baseline conditions. *Journal of Homeland Security and Emergency Management, 7*(1). https://doi.org/10.2202/1547-7355.1732

Devers, K. J., Lallemand, N. C., Burton, R., Kahwati, L., Mccall, N., & Zuckerman, S. (2013). *Using qualitative comparative analysis (QCA) to study patient-centered medical homes* (Issue September).

Di Gregorio, L. T., & Soares, C. A. P. (2017). Post-disaster housing recovery guidelines for development countries based on experiences in the American continent. *International Journal of Disaster Risk Reduction, 24*(2017), 340–347. https://doi.org/10.1016/j.ijdrr.2017.06.027

Du, F., Okazaki, K., Ochiai, C., & Kobayashi, H. (2016). Post-disaster building repair and retrofit in a disaster-prone historical village in China: A case study in Shangli, Sichuan. *International Journal of Disaster Risk Reduction, 16*(2016), 142–157. https://doi.org/10.1016/j.ijdrr.2016.02.007

Edgington, D. W. (2011). Reconstruction after natural disasters: The opportunities and constraints facing our cities. *Town Planning Review, 82*(6). https://doi.org/10.3828/tpr.2011.35

Griffey, A. M. (2010). *Infrastructure disaster management : Insight from small business leaders after a mega disaster* (Issue December). University of Phoenix.

Han, Z. (2014). *From vulnerability to resilience : Long-term livelihood recovery in rural china after the 2008 Wenchuan earthquake*. University of Delaware.

He, L. (2019). Identifying local needs for post-disaster recovery in Nepal. *World Development, 118*(2019), 52–62. https://doi.org/10.1016/j.worlddev.2019.02.005

Hu, H., Lei, T., Hu, J., Zhang, S., & Kavan, P. (2018). Disaster-mitigating and general innovative responses to climate disasters: Evidence from modern and historical China. *International Journal of Disaster Risk Reduction, 28*, 664–673. https://doi.org/10.1016/j.ijdrr.2018.01.022

Jordan, E. (2012). *Pathways to community recovery: A qualitative comparative analysis of post-disaster outcomes*. University of Colorado.

Li, C., Wang, M., & Liu, K. (2018). A decadal evolution of landslides and debris flows after the Wenchuan earthquake. *Geomorphology, 323*, 1–12 (October 2005). https://doi.org/10.1016/j.geomorph.2018.09.010

Li, X., Lam, N., Qiang, Y., Li, K., Yin, L., Liu, S., & Zheng, W. (2016). Measuring county resilience after the 2008 Wenchuan earthquake. *International Journal of Disaster Risk Science, 7*(4), 393–412. https://doi.org/10.1007/s13753-016-0109-2

Lizarralde, G. (2004). *Organisational system and performance of post-disaster reconstruction projects*. Universite de Montreal.

Lu, Y., & Xu, J. (2015). Comparative study on the key issues of postearthquake recovery and reconstruction planning: Lessons from the United States, Japan, Iran, and China. *Natural Hazards Review, 16*(3), 04014033. https://doi.org/10.1061/(asce)nh.1527-6996.0000172

Nejat, A., & Ghosh, S. (2016). LASSO model of postdisaster housing recovery: Case study of hurricane sandy. *Natural Hazards Review, 17*(3), 04016007–1–13. https://doi.org/10.1061/(asce)nh.1527-6996.0000223

Olshansky, R. B. (2006). Planning after hurricane Katrina. *Journal of the American Planning Association, 72*(2), 147–153. https://doi.org/10.1080/01944360608976735

Okamoto, K., & Ishikawa, Y. (2013). Traditional Wisdom and Modern Knowledge for the Earth's Future. In J. Noritaka Yagasaki (Ed.), *International Geographical Union Kyoto Regional Conference*. Sprnger.

Platt, S. (2018). Factors affecting the speed and quality of post-disaster recovery and resilience. In S. Ó. Rajesh Rupakhety (Ed.), *Earthquake engineering and structural dynamics in memory of Ragnar Sigbjörnsson* (pp. 369–403). Springer. https://doi.org/10.1007/978-3-319-62099-2

Rani, W. N. M. W. M., Nifa, F. A. A., Ismail, M. N., & Khalid, K. N. (2017). Planning for post disaster recovery: Lesson learnt from flood events in Kelantan Malaysia. *AIP Conference Proceedings, 1891*(1), 0201431–1–6. https://doi.org/10.1063/1.5005476

Rihoux, B. (2006). Qualitative comparative analysis (QCA) and related systematic comparative methods: Recent advances and remaining challenges for social science research. *International Sociology, 21*(5), 679–706. https://doi.org/10.1177/0268580906067836

Roosli, R., & O'Keefe, P. (2013). Post-disaster housing and management in Malaysia: A literature review. *International Journal of Disaster Resilience in the Built Environment, 4*(2), 168–181. https://doi.org/10.1108/IJDRBE-06-2011-0022

Rose, A. (2011). Resilience and sustainability in the face of disasters. *Environmental Innovation and Societal Transitions, 1*(2011), 96–100. https://doi.org/10.1016/j.eist.2011.04.003

Sadiqi, Z., Trigunarsyah, B., & Coffey, V. (2017). A framework for community participation in post-disaster housing reconstruction projects: A case of Afghanistan. *International Journal of Project Management, 35*(5), 900–912. https://doi.org/10.1016/j.ijproman.2016.11.008

Silva, J. da. (2010). Lessons from Aceh. Key considerations in post-disaster reconstruction. In *Post-disaster reconstruction: Lessons from Aceh*. Practical Action Publishing. https://doi.org/10.4324/9781849775137

Tauber, G. (2015). Architects and rural post-disaster housing: Lessons from South India. *International Journal of Disaster Resilience in the Built Environment, 6*(2), 206–224. https://doi.org/10.1108/IJDRBE-07-2013-0025

Tucker, S., Gamage, A., & Wijeyesekera, C. (2014). Some design aspects of sustainable post-disaster housing. *International Journal of Disaster Resilience in the Built Environment, 5*(2), 163–181. https://doi.org/10.1108/IJDRBE-06-2012-0019

Xu, J., & Lu, Y. (2018). Towards an earthquake-resilient world: From post-disaster reconstruction to pre-disaster prevention. *Environmental Hazards, 17*(4), 269–275. https://doi.org/10.1080/17477891.2018.1500878

Xu, P., Lu, X., Zuo, K., & Zhang, H. (2014). Post-Wenchuan earthquake reconstruction and development in China. In K. T. L. Naim Kapucu (Ed.), *Disaster and development examining global issues and cases*. Springer. https://doi.org/10.1007/978-3-319-04468-2_25

Shao, Y. (2017). Urban resilience in China's post-disaster reconstruction planning: The cases of Weizhou and Yingxiu towns, Wenchuan county. *The Chinese University of Hong Kong, 4*.

Zhang, Y., & Peacock, W. G. (2010). Planning for housing recovery? Lessons learned from hurricane Andrew.. *Journal of the American Planning Association, 76*(1), 5–24. https://doi.org/10.1080/01944360903294556

Chapter 6
Pathways to the Success/Failure of Recovery Projects Based on Communities' Perspective; Bam Recovery Project

Abstract Although recent studies in recent literature have provided explanations for the causes of success and failure in recovery projects following socio-natural disasters, there is a need for a concise understanding of how different combinations of factors may contribute to recovery failure or success. In this chapter, to examine the community recovery pathways after the 2003 Bam earth-quake, the researchers conducted a fuzzy-set qualitative comparative analysis of 11 neighbourhoods in Bam city and Baravat (the neighbourhoods represent the division of the areas damaged by the earthquake, as presented by local government). The success of socio-natural disaster recovery projects is presented in three pathways in which the absence or presence of public engagement significantly influenced the results. The results indicate that a recovery project should respond to the project's needs within the project's continuous life cycle. Additionally, public participation and prompt rescue operations significantly affect project success. On the other hand, neglecting the needs of people and the area's climate in housing design has led to project failure. Moreover, correlation coefficient analysis among CSFs revealed how these factors might affect each other. It is expected that the findings from this study can be used to develop strategies for empowering people in recovery projects and to develop housing guidelines that respect residents' needs while focusing on on-time and sufficient rescue processes.

Keywords Disaster recovery success · Public perspective · Bam earthquake

6.1 Application of Fuzzy-Set QCA Analysis

Qualitative Comparative Analysis can bridge gaps between qualitative and quantitative data analysis techniques. Quantitative data analysis methods evaluating a large sample size restrict the researchers to the subsequent level of information on the cases. However, QCA requires a deep understanding of the selected cases as the final quantitative results will be taken back within the cases to the interpretation stage based on the in-depth knowledge of case studies. Therefore, the researcher must

H. Bahmani et al., *Socio-Natural Disaster Recovery Projects*, Resilient and Friendly Living Environments, https://doi.org/10.1007/978-981-95-0113-7_6

keep intimacy with cases. QCA also compares more complex causalities than quantitative methods, which primarily assess the impact of single factors on the outcome. By taking advantage of quantitative data analysis methods, set theory, and Boolean algebra, QCA can also answer qualitative data analysis methods' deficiencies (Ragin, 2008). The QCA usually initiates by modelling causes and effects and continues by analyzing data through a back-and-forth procedure while keeping a dialogue with the cases. Since this procedure requires the researcher to repeat and revise some of the previous steps, Table 6.1 briefly discusses the steps and their subsequent section in this book. In the equations in that table, X_i and Y_i are membership scores of causal condition and outcome of case i, respectively (Jordan, 2012).

This chapter modified the KPIs, CSFs, and their interaction model (Chapter 5) to assess the recovery per the communities' perspective. The success/ failure pathways for the Bam recovery project have been identified through a careful proposal of success dimensions and critical success factors. However, since those identified factors delve into looking at SNDR projects as a whole and engaging multi-level stakeholders (government, NGOs, public), evaluating all the identified factors is a taught work that requires access to more resources and time. On the other hand, considering the significance of public satisfaction with the recovery outcome and its direct and indirect impacts on the recognized sub-systems of SNDR projects (Chapter 3), this research has studied recovery success pathways based on the community perspective. Therefore, the authors have modified the suggested success dimensions (outcomes) and their influential factors (conditions) to evaluate how the combination of conditions led to the success/ failure of recovery outcomes per households' perspective. The calculations from steps 4–6 were conducted through fsQCA software. It is necessary to mention that, in complex causalities analysis, which is asymmetry, the negations of outcomes have to be modelled separately since the negation of the recognized solution(s) may not result in the non-occurrence of the outcome(s). Therefore, the authors also modelled the negation of each outcome separately, resulting in eight models with four outcomes and their negations.

6.2 Evaluating Recovery Success Per Communities' Perspective

The research identified KPIs and CSFs for SNDR projects (Chapter 5), which involve the "public" as one of the primary beneficiaries. Therefore, the researchers applied the QCA method to figure out how the combination(s) of different conditions might lead to a successful SNDR project from the perspective of communities. The QCA, which can provide a middle solution between qualitative and quantitative methods, has been chosen as an appropriate analysis method. It can consider the combination of parameters affecting specific outcomes and recognize the most possible scenario leading to successful recovery.

Table 6.1 Fuzzy-set QCA steps in this study

Steps		Tools	Data analysis type	Key features and formulas	
1	Model generation	Theoretical knowledge	Qualitative	Conditions Outcomes	
2	Case selection	Selection of cases with homogeneous backgrounds and heterogenous features	Qualitative	NOT random case selection	
3	Calibration & data matrix	1. Case knowledge 2. Fuzzy set 3. Set theory	Qualitative & Quantitative	Set theory	1. Average score = All factors are equally important 2. Max. score = One 4factor must be 5present 3. Min. score = All factors must be present
				Fuzzy set	Scores between 0–1. Use of continuous or N-value fuzzy set per type of information
4	Truth table	Set theory	Quantitative	Set theory	* Logical AND + Logical OR ~ Logical NOT
				If sufficiency $\geq 0.85 - \gg$ input for logical minimization	$\mathit{Necessity of\ conditions} = \sum \frac{(\min(XiYi))}{\sum Yi}$ (1) $\mathit{Sufficiency of\ conditions} = \sum \frac{(\min(XiYi))}{\sum Xi}$ (2)

(continued)

Table 6.1 (continued)

Steps		Tools	Data analysis type	Key features and formulas
5	Logical minimization	1. Quine-McCluskey algorithm (Tabular method) 2. Solution algorithms (This study has applied intermediate solution.)	Quantitative	1. Recognizing prime implicants 2. Finding minimal formula
6	Interpretation	1. Minimal formula 2. Set theory 3. Case knowledge	Quantitative & Qualitative	$Rowconsistencyof\ confiration = \sum \frac{(\min(XiYi))}{\sum Yi}$ (3) $Coverageof\ configuration = \sum \frac{(\min(XiYi))}{\sum Xi}$ (4)
				The minimal formula must be interpreted by going back to cases

Furthermore, the QCA model showing the interactions among the conditions and outcomes has been adjusted in Fig. 6.1, and the rest of the factors, which are not directly linked to the community perspective, have been accessed previously using NCA in Chapter 6. The established conditions and outcomes were revised based on knowledge of the neighbourhoods. The domain factors such as evacuation facilities, food availability, and equity among the populations were removed from the analysis as they did not vary across the neighbourhood (Ragin, 2009). The study placed the recognized factors in a tree-shaped categorization to reduce the number of parameters and modelling complexity (Befani, 2016). The following framework investigated how the combination of recognized conditions may affect residents' long-lasting and holistic recovery. The final score of the ultimate goal (main objective) was calculated through a weighted average of the mean scores for *outcomes* A_1, A_2, and A_3, giving twice the weight to outcome A_3 based on the Delphi result in Chapter 4.

The data were collected through structured interviews with households in 11 Bam city and Baravat neighbourhoods. Since Bam was reconstructed by dividing it into several districts assisted by different provinces, the neighbourhoods were selected according to the damage zoning map after the earthquake. The researchers attempted to cover various sections governed by each of the three municipalities in Bam. Moreover, Baravat was chosen because of its proximity to Bam city (almost 10 km) and the high level of destruction caused by the earthquake. Figure 6.2 shows the selected neighbourhoods (coloured districts) where the public survey was conducted.

Two trained interviewers conducted the interviews in chosen neighbourhoods during August 2021. The interviewers have been provided with a detailed guide for conducting public interviews (general requirement to interact with people), selecting interviewees, and research goals and backgrounds. The training was in the form of written materials and online discussions. The researchers have kept close contact with them during the fieldwork and held online meetings once a day. In case of any difficulty during the fieldwork, they immediately contacted the authors and asked for further information. The main criterion for choosing the interviewees was their living duration in the damaged area; interviewees must have lived in Bam or Baravat before the earthquake. Moreover, the interviewers selected only one person from each household to avoid redundant data. The fieldwork lasted for five days and resulted in 122 interviews with members of the public, among which 67% of interviewees were male, 48% were between 31–40 years old, and 45% had been working in government-related organizations/agencies. An integrated online form was utilized to collect the responses. The interview responses were recorded in Farsi and then stored. The interviewers also observed the neighbourhoods, city facilities, urban development, and emergency preparedness in the selected neighbourhoods. Finally, the non-weighted average of each neighbourhood's responses represented the neighbourhood's score. In the case of significant differences among the collected responses in one neighbourhood, the researchers revised the data based on the observations and secondary documents. Figure 6.3 shows Bam's ancient citadel over time.

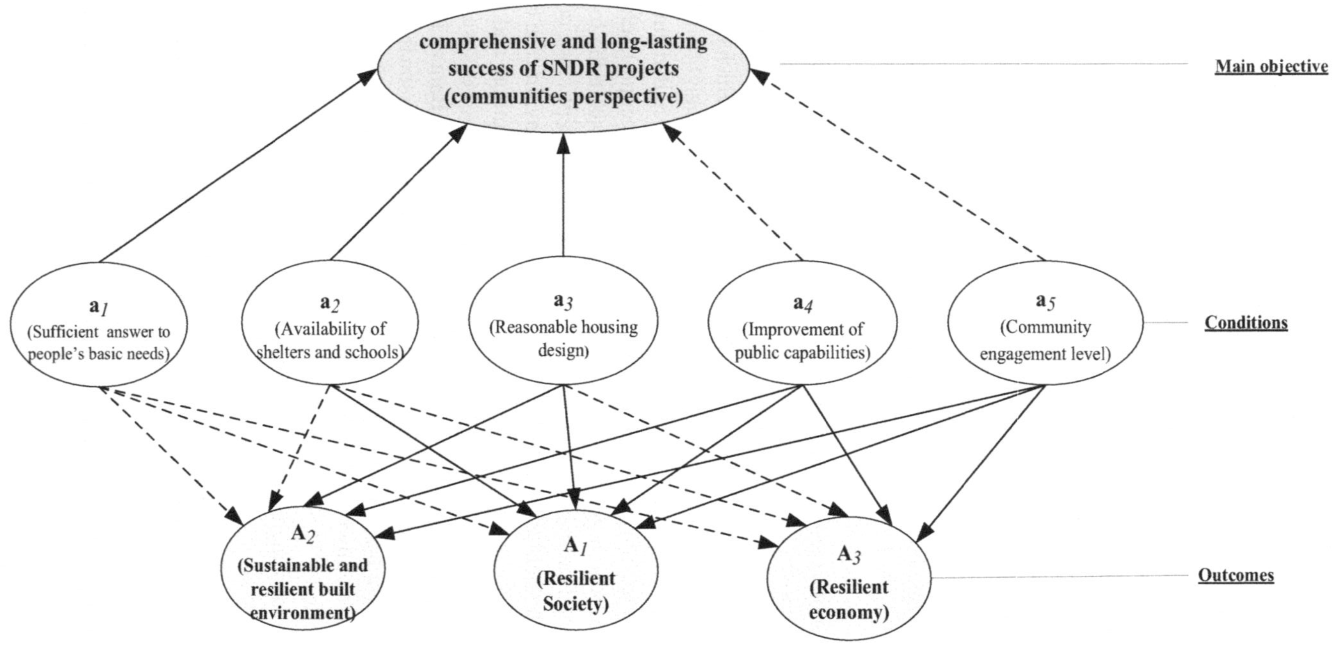

Fig. 6.1 Interactions between KPIs and CSFs to assess success per communities' perspective. Dashed lines indicate that the condition's presence and absence may affect the connected outcome, while the undashed lines represent the presence of the condition forming the subsequent outcome

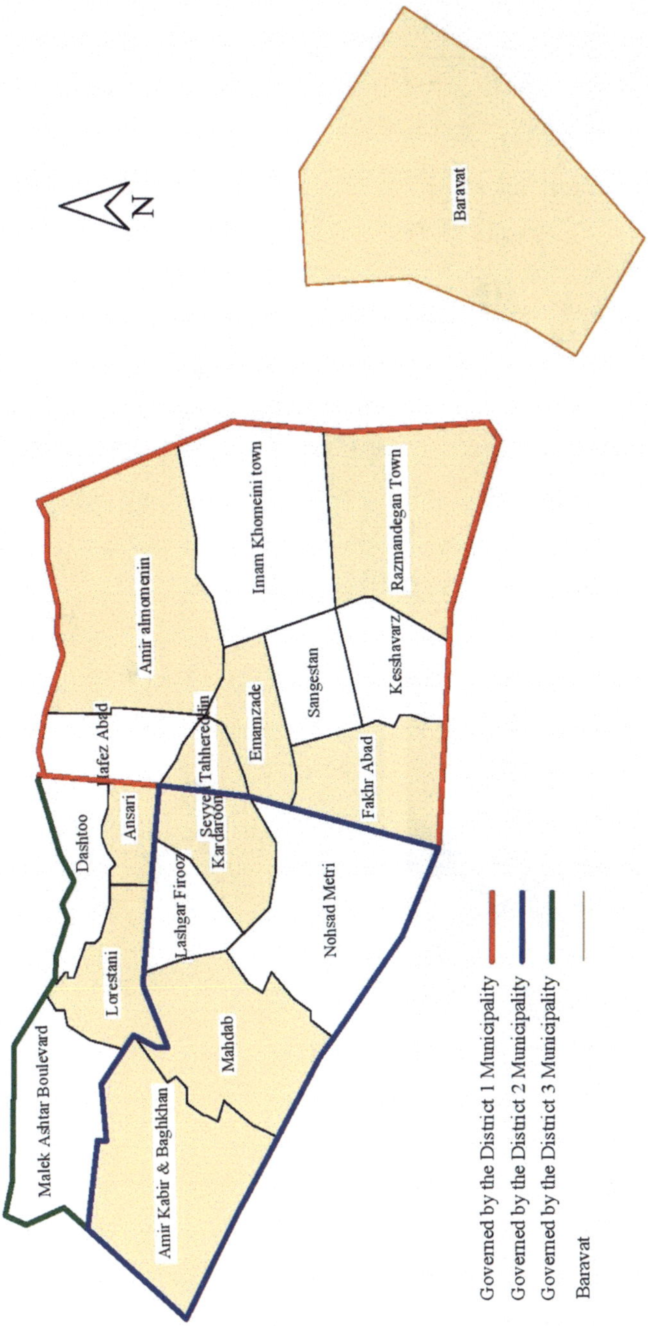

Fig. 6.2 Selected neighbourhoods in Bam and Baravat

Fig. 6.3 Arg-e-Bam over time (top-left photo) before the earthquake, (top-right photo) after the earthquake, and (photo below) now

6.3 Assigning Set Membership Scores for CSFs and KPIs

Once the data were collected, the researchers began the data analysis by initiating data calibration; all the collected data were transformed into quantitative data based on set theory to assign the set membership scores for each case. The answers to the questions were calibrated using fuzzy-set logic and set theory, referring to the researcher's general knowledge about the neighbourhoods. Since most factors consist of several subfactors, calculation formulas depend on the role and importance of the factors. If all the subfactors are equally important, the average score of the subfactors is calculated. For example, four subfactors were equally important for outcome A_1, i.e., resilient society. Furthermore, the total score for this outcome was a non-weighted average of the four subfactors. If previous research and knowledge of neighbourhoods emphasized one subfactor, the average score was calculated considering weights. For example, reasonable housing design emphasizes consideration of the culture and local needs. Therefore, the average score was calculated, giving twice the weight to housing based on the culture of the damaged area and the needs of the people who were affected, and one weight was given to climate design. If the researcher recognized all the factors must be present in the set, the minimum score was calculated. For example, "roads' improvement" is a sub-outcome of outcome A_2, which consists of several subfactors. Since the authors recognized that all the factors must be present to have better roads, the minimum score of the subfactors formed the total score of this sub-outcome. The maximum score in the calibration table indicates that the presence of one of the parameters is enough for the neighbourhood to be considered in the set. Tables 6.2 and 6.3 represents the conditions and outcomes calibration rubrics, respectively.

6.3.1 Calibration Rules of CSFs (Conditions)

(1). Sufficient Emergency Answer to Basic Public Needs (a_1)

Providing appropriate answers for the short-term needs of residents can greatly affect the success of the SNDR project from the communities' perspective. Emergency public needs can be grouped into different categories: sanitation, food, and saving lives. In the aftermath of the Bam earthquake, as the Iranian Red Cross has done the food distribution among the suffered people, it has been excluded from the cases as all neighbourhoods showed the same results (domain condition). Among the two factors of this condition, sanitation (b_1) includes child parameters facilitating the data collection and analysis. The water availability in b1 has gained twice weight on average as it has been identified to be more critical based on the Bam recovery project.

(2). Availability of Shelters and Schools (a_2)

Safe shelters are necessary for saving lives and have gained consensus in the Delphi survey. However, the importance of education is somehow neglected in

Table 6.2 Conditions' calibration rubrics

<table>
<tr><td rowspan="4">a_1 (Sufficient emergency answer to basic public needs)</td><td colspan="4">$a_1 = (b1, 2{*}b2)/3$</td></tr>
<tr><td colspan="2">b1, Sanitation level</td><td colspan="2">b2, Rescue and site investigation</td></tr>
<tr><td>(b1-1) Water availability in the shelters:
[0] No access to public toilets and difficulty accessing water
[0.33] Access to public toilets but difficulty accessing water
[0.67] No access to public toilets, but access to water
[1] Access to both public toilets and water</td><td>(b1-2) Medical service availability in the shelters:
[0] No access
[0.5] Sometimes available
[1] Always available</td><td colspan="2">(b2-1) Time of first rescue team arrival:
[0] One week after the earthquake
[0.4] 4–7 days after the earthquake
[0.8] Within three days after the earthquake
[1] Next day after the earthquake</td></tr>
<tr><td colspan="2">$b1 = (2{*}b1\text{-}1 + b1\text{-}2)/3$</td><td colspan="2">$b2 = \sum (b2\text{-}1)$</td></tr>
<tr><td rowspan="4">a_2 (Availability of shelters and schools)</td><td colspan="4">$a_2 = (b3 + 2{*}b4)/3$</td></tr>
<tr><td>b3, Education availability</td><td colspan="3">b4, Shelter availability</td></tr>
<tr><td>(b3-1) Teachers availability after the earthquake:
1 Volunteer
[1] Governmental organization (ministry of education)
[1] Non-governmental organization
[0] No access to teachers</td><td>(b4-1) Waiting time to find emergency shelter:
[0] More than one month
[0.2] 2–4 weeks
[0.4] 1–2 weeks
[0.7] One week
[1] Just after the earthquake</td><td>(b4-2) Duration of living in emergency shelters:
[0] More than 2 years
[0.2] 1–2 years
[0.4] 6 months to 1 year
[0.7] 2–6 months
[1] Less than 2 months</td><td>(b4-3) Duration of living in transitional homes:
[0] More than 2 years
[0.2] 1–2 years
[0.4] 6 months to 1 year
[0.7] 2–6 months
[1] Less than 2 months</td></tr>
<tr><td>$b3 = \sum (b3\text{-}1)$</td><td colspan="3">$b3 = (4\text{–}1 + b4\text{-}2 + b4\text{-}3)/3$</td></tr>
<tr><td rowspan="2">a_3 (Reasonable housing design)</td><td colspan="4">$a_3 = (2{*}b5 + b6)/3$</td></tr>
<tr><td colspan="2">b5, housing based on local culture and needs</td><td colspan="2">b6, Climate consideration in housing design</td></tr>
</table>

(continued)

Table 6.2 (continued)

	(*b5-1*) Living experience, preference of old or new house: [0] The old house [0.5] Both are the same [1] New house	**(*b5-2*) House layout modifications:** [0] Any change [1] No change	**(*b6-1*) Indoor thermal comfort, Winter:** [0] Freezing, intolerable [0.5] Moderate, tolerable [1] Warm, desirable	**(*b6-2*) Indoor thermal comfort, summer:** [0] Scorching, intolerable [0.5] Moderate, tolerable [1] Cool, desirable
$b5 = Maximum. (b5\text{-}1, b5\text{-}2)$			$b6 = (b6\text{-}1 + b6\text{-}2)/2$	
a_4 (Improvement of public capabilities)	$a_4 = Minimum. (b7, b8)$			
	b7, Community training		*b8*, Livelihood development	
	(*b7-1*) Training availability: [1] Emergency aids [1] Rescuing and self-protection [1] Safe construction [1] Others [0] No	**(*b7-2*) Training time:** [0] No training [0.5] After the earthquake [1] Before the earthquake	**(*b8-1*) Received financial aid:** [1] House rebuild/repair grant [1] Bank loans for housing [1] Subsidy/ grant for business [1] Bank loans for business [1] Monthly stipend [1] Other [0] Nothing	**(*b8-2*) Duration of financial aid:** [0] No financial aids [0.33] Less than 6 months [0.67] 6 months to 1 year [1] Till the end of project
	$b7 = (b7\text{-}1 + b7\text{-}2)/2$		$b8 = (b8\text{-}1 + b8\text{-}2)/2$	
a_5 (Community engagement level)	$a_5 = (b9 + 2{*}b10)/3$			
	b9, Public participation in decision-making		*b10*, Public participation in construction	

(continued)

Table 6.2 (continued)

	***(b9-1)* Public participation in site selection:** [0] Relocated to the new site [1] Settled in the old site	***(b9-2)* Keeping public informed:** [1] By Shora (community centers) [1] By local news [1] By Setads [1] By volunteers/ NGOs [1] By neighbours/ friends [1] Other [0] NO shared information	***(b10-1)* Public participation in planning:** [0] Choosing one of the portfolio plans for house layout [1] Participation in housing planning with the consultant's assistance	***(b10-2)* Public participation in reconstruction:** [0] Choosing the contractor and supervision on his work [1] Self-construction as a contractor	***(b10-3)* Public participation in procurement:** [0] Contractor in charge of purchasing materials [1] Householder in charge of purchasing materials
	b9 = + (*b9-1* + *b9-2*)/2		*b10* = (*b10-1* + *b10-2* + *b10-3*)/3		

Table 6.3 Outcomes' calibration rubrics

A_1 (Resilient Society)	$A_1 = (B1 + B2 + B3 + B4)/4$				
	B1, Social connection	*B2*, Psychological support	*B3*, safety and security	*B4*, Life satisfaction	
	(*B1-1*) Living with the same family and neighbours after the earthquake: [0] Loss of family members due to death or immigration (earthquake) [0.5] There are new members in the community or family [1] All the social connections are the same as before the earthquake	**(*B2-1*) Availability of psychological consultation after the earthquake:** [1] By volunteers [1] By Red Crescent [1] By friend/ relatives [1] By NGOs [0] No consultation	**(*B3-1*) Experience of robbery/ crime after the earthquake:** [0] Repeatedly [0.2] Often [0.4] Sometimes [0.7] Few times [1] Rarely	**(*B4-1*) Living experience, preference of old or new house:** [0] The old house 0 [0.5] Both are the same [1] New house	**(*B4-2*) City's major changes:** [0] Decline in industry [1] Boomed industry [0] Decline in agriculture [1] Boomed agriculture [0] Decline in tourism [1] Boomed tourism [0] Polluted environment [1] Cleaner environment [0] Worse transportation system [1] Better transportation system [0] Worse education, health system [1] Better education, health system [0] Unsafe and uncomfortable houses [1] Safer and better houses
	$B1 = \sum (B1\text{-}1)$	$B2 = \sum (B2\text{-}1)$	$B3 = \sum (B3\text{-}1)$	$B4 = Maximize. (B4\text{-}1, B4\text{-}2)$	
A_2 (Sustainable and resilient built environment)	$A_2 = (B5 + 2{*}B6 + 2{*}B7 + B8 + B9)/7$				
	B5, Population per capita in houses		*B6*, Hygiene water accessibility	*B7*, Electricity accessibility	

(continued)

Table 6.3 (continued)

	(*B5-1*) Changes of the available housing area for each person before and after the earthquake: [0] Decrease in the available housing area per person [1] Increase in the available housing area per person		**(*B6-1*) Resuming time of the water availability in the neighbourhood:** [0] More than two years [0.4] One- two year [0.8] 6 months to one year [1] Less than 6 months	**(*B7-1*) Resuming time of the electricity availability in the neighbourhood:** [0] More than two years [0.4] One- two year [0.8] 6 months to one year [1] Less than 6 months	
	$B5 = \sum (B5\text{-}1)$		$B6 = \sum (B6\text{-}1)$	$B7 = \sum (B7\text{-}1)$	
	B8, Roads improvement			*B9*, Education and hospitals' improvement	
	(*B8-1*) Resuming time of the roads in the neighbourhood: [0] Some roads have not been finished yet [0.2] More than three years [0.4] Two–three years [0.7] One- two year [1] Less than one year	**(*B8-2*) Roads' quality:** [0] Extremely worse now [0.2] Worse now [0.4] No change [0.7] Better now [1] Extremely better now	**(*B8-3*) Accessibility to the Province center and nearby cities:** [0] It is much difficult now [0.5] It has not changed [1] It is much easier now	**(*B9-1*) Education quality:** [0] Extremely worse now [0.2] Worse now [0.4] No change [0.7] Better now [1] Extremely better now	**(*B9-2*) Medical service quality:** [0] Extremely worse now [0.2] Worse now [0.4] No change [0.7] Better now [1] Extremely better now
	B8 = Minimum. (B8-1, B8-2, B8-3)			$B9 = (B9\text{-}1 + B9\text{-}2)/2$	
A_3 (Resilient economy)	$A_3 = (B10 + B11 + 2{*}B12)/4$				

(continued)

Table 6.3 (continued)

	B10, Business recovery			*B11*, Employment rate		*B12*, Household income
	(*B10-1*) Old businesses recovery: [0] No re-operation of the old businesses [0.5] Partial recovery of the old businesses [1] Full recovery of the old businesses	**(*B10-2*) Business growth rate (before and after the earthquake):** [0] Declined [0.5] No change [1] Increase	**(*B10-3*) Job stability in the city:** [0] People with changed career [1] People with the same career	**(*B11-1*) Employment status:** [0] Employee [1] Business owner	**(*B11-2*) Working place during reconstruction:** [0] Other places [1] Bam	**(*B12-2*) Households' income change before and after the earthquake:** [0] better income before the earthquake [0.5] No change in income [1] Better income after the earthquake
	$B10 = (B10\text{-}1 + B10\text{-}2 + B10\text{-}3)/3$			$B11 = (B11\text{-}1 + B11\text{-}2)/2$		$B12 = \sum (B12\text{-}1)$

the literature, and the Delphi panel did not agree upon its importance. Therefore, as both factors must be present to be included in the set, the average score, giving the twice value to the availability of the shelter, forms this condition's score. Moreover, the availability of education (b_3) and the organization in charge are scored equally for all the organizations unless no teacher was provided.

(3). Reasonable Housing Design (a_3)

From the perspective of households, suitable housing can be seen within two groups: cultural design and climate design. However, most of the residents have emphasized the significance of the cultural design of houses. The inappropriate house layout was one of the primary reasons causing the residents' dissatisfaction. Moreover, the literature and Delphi survey findings reveal that public needs may have more impact on reasonable housing design. Furthermore, the average cultural and climate design score, twice the weight for cultural design, is used to calculate this condition's score.

(4). **Improvement of public capabilities to cope with disasters (a_4)**

Public training and its conduction time, the types of received financial aid, and their payment duration have been considered the sub-factors of this condition. Any training and financial assistance were equally scored, while training conduction time and duration of the financial aids were calibrated within three and four fuzzy set-range scores, respectively. As livelihood development and enhancement in public capabilities should be present to consider the case inside the set, the minimum score of the parameters represents this condition's total score.

(5). Community Engagement Level (a_5)

The community engagement level has been evaluated, considering two phases: public engagement in decision-making and construction. As people's participation in the former phase may interfere with the project's technical planning (Flores Salas, 2016), twice the weight has been given to public involvement in the construction phase, including people's engagement in planning, reconstruction, and procurement. Any information dissemination approach scored equally, and a simple average was calculated for the decision-making stage. The same method has measured public participation in the construction phase while dividing and evaluating people's engagement levels in different topics, such as planning, reconstruction, and procurement.

6.3.2 *Calibration Rules for KPIs (Outcomes)*

(1). Resilient Society (A_1)

Society's resilience has been measured by short-term, mid-term, and long-term sub-factors. While revitalizing the social ties assessed the survivors' connections by a three-ranged fuzzy score, the availability of the psychological consultations conducted by any organization has been equally ranked. The experience of robbery or crime and the residents' housing preference were the factors that have been appraised by five and three-ranged fuzzy scores, respectively. However, the significant changes in the city from the perspective of the residents have utilized crisp set values (0,1). As the four factors are considered to have equal impacts on this outcome, the non-weighed average of their scores represents the total score of this outcome.

(2). Sustainable and Resilient Built Environment (A_2)

Given twice the score to the water and electricity availability as the agreed factors from the Delphi survey (basic community needs), the sustainability and resilience of the built environment have been measured by calculating the average of five sub-factors. The growth rate in the population per capita before and after the earthquake uses crisp values (0 for increasing the value in recent years, 1 for growing). Resuming the time of access to water and electricity was evaluated by four-ranged fuzzy scores. Roads' quality improvement, restarting time, and accessibility to the other cities, alongside improved education and medical service, assessed the other two sub-factors. While a minimum of the child parameters of road improvement has formed this sub-factor value, a non-weighted educational and medical facilities improvement score has been calculated for education and hospitals' improvement.

(3.) Resilient Economy (A_3)

The average of three sub-factor, namely, business recovery, employment rate, and household income, giving twice the weight to household income, has evaluated the resilient economy. Due to Iran's unstable financial status and currency fluctuation in recent years, the researchers have asked the respondents to rank their income improvement compared to before the earthquake.

6.4 Departure for Structuring Success and Failure Pathways

After the data were calibrated, a data matrix containing the scores for each condition and outcome for each neighbourhood was constructed (Table 6.4). The "necessity" and "sufficiency" of individual conditions can also help the researchers to input conditions that are logically connected to the outcomes for the next phase. While necessity shows the degree to which an outcome is a subset of causal conditions, sufficiency reveals the degree to which the causal condition is a subset of the outcome(s) (Befani, 2016; Ragin, 2009). The necessity and sufficiency are calculated by formulas that appear in Table 6.1.

The analysis has continued by structuring truth tables. Truth Table shows "*the logically possible combinations of causal conditions (number of causal combinations* $= 2^k$*, where k is the number of causal conditions) and the empirical outcome associated with each configuration*" (Ragin, 2008). Since truth table construction is the most important part of the analysis procedure, Table 6.5 presents some configurations of the study's truth table (The complete truth table for each outcome has 32 rows). First, the membership score of each case for the recognized configurations was

Table 6.4 Data matrix representing fuzzy scores for cases

Case Num	Case name	a_1	a_2	a_3	a_4	a_5	A_1	A_2	A_2	Ultimate
1	Amir Kabir & Baghkhan	0.820	0.436	0.49	0.729	0.522	0.43	0.44	0.42	0.428
2	Mahdab	0.724	0.286	0.426	0.528	0.556	0.37	0.42	0.38	0.386
3	Fakhr Abad	0.750	0.642	0.509	0.881	0.411	0.61	0.44	0.47	0.498
4	Razmandegan town	0.716	0.411	0.871	0.498	0.556	0.44	0.47	0.38	0.418
5	Emamzadeh	0.814	0.665	0.474	0.539	0.444	0.47	0.43	0.42	0.436
6	Amir-al-momemnin	0.593	0.805	0.398	0.639	0.389	0.63	0.28	0.34	0.396
7	Lorestaniha	0.575	0.480	0.432	0.432	0.647	0.49	0.47	0.5	0.488
8	Koradoon	0.667	0.676	0.492	0.799	0.520	0.34	0.43	0.59	0.488
9	Ansari	0.477	0.580	0.409	0.250	0.791	0.52	0.54	0.46	0.494
10	Baravat	0.650	0.640	0.492	0.796	0.556	0.44	0.34	0.39	0.390
11	Seyyed Taher	0.651	0.589	0.444	0.527	0.417	0.32	0.46	0.41	0.400

calculated by taking the minimum score of the individual conditions corresponding to that case. Subsequently, after calculating the raw consistency of causal configurations and considering a threshold value of more than 0.92, the researchers decided whether or not that configuration represents the outcome using binary numbers. Finally, cases representing the configurations have been identified. Figure 6.4 graphically compared the neighborhood's conditions and outcomes' scores.

Next, the truth table was accessed through logical minimization, illustrating the pathways (combinations of condition(s) leading to a defined outcome). "Consistency" and "coverage" are two measurements used to evaluate pathways recognized as "parameters of the fit". The degree to which the given cases of the causal conditions present outcomes can be calculated by consistency, the same as necessity calculation. Coverage indicates the degree to which the pathway was observed among the cases, similar to the sufficiency formula. Usually, the researchers prefer a consistency of at least 0.8 among the recognized pathways (Befani, 2016). The parameters of fit should be calculated for each pathway in the truth table by use of the equation in Table 6.1. In this study, the ultimate outcome, the comprehensive and long-lasting success of SNDR projects per community's perspective, was evaluated by calculating the average scores of the major outcomes, i.e., A_1, A_2, and A_3, given twice the weight of A_3, following the Delphi results. The next paragraphs will explain the calculation procedure done by fsQCA software for outcome A_1 as an example.

(1). The calculation has continued by computing the necessity and sufficiency of each condition to reach the outcome (Table 6.6).

As no strong theoretical background between emergency response to people's basic needs and resilient society has been found, the model has considered this condition's absence and presence (a_1). Therefore, using $\sim a_4$ as the prime implicant, the below intermediate solution has resulted from the truth table.

Table 6.5 Truth table for outcome A_1

Possible configurations	Cases' membership score (*n* represents case number)						Row consistency	Outcome representative? (0 = no, 1 = yes)	Cases
	Case 1		Case 2	Case 3	…	Case 11			
a_1*a_2* a_3* a_4* a_5	min ($a_{1,1},\ldots, a_{i,n}$)	0.436	0.286	0.411	…	0.444	0.79	0	LR[1]
a_1*a_2* ~ a_3* a_4* a_5		0.436	0.286	0.436	…	0.527	0.91	0	8,10
a_1*a_2* a_3* a_4* ~ a_5		0.436	0.286	0.436	…	0.444	0.94	1	3
a_1*a_2* ~ a_3* a_4* ~ a_5		0.436	0.286	0.491	…	0.527	0.91	0	5,6,11
a_1* ~ a_2* ~ a_3* a_4*a_5		0.51	0.528	0.358	…	0.411	0.92	1	1,2
…		…	…		…	…	…	…	…
a_i (*i* represents conditions)	Transposed data matrix	0.820	0.742	0.75	…	0.522			
a_2		0.436	0.286	0.642	…	0.556			
a_3		0.49	0.426	0.509	…	0.444			
a_4		0.729	0.528	0.498	…	0.527			
a_5		0.522	0.556	0.411	…	0.417			
Outcome A_1		0.43	0.43	0.61	…	0.32			

[1] LR: logical reminder. This causal configuration is logical reminder, since there is no case representing this causal configuration. Logical remiders can be used in order to simplifying the complex solutions; howver, using logical remiders depends on presence of strong evidence for their impact on the outcome.

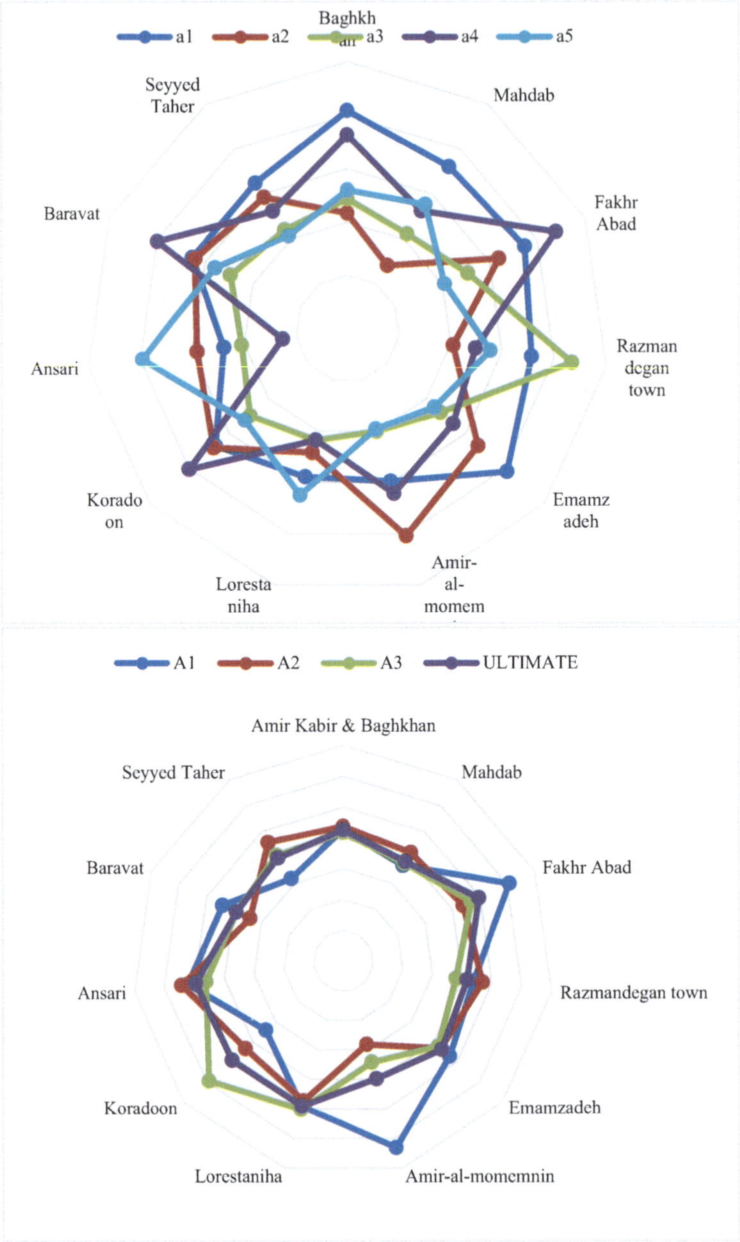

Fig. 6.4 Spider diagrams of conditions and outcomes in neighbourhoods

Table 6.6 Necessity and sufficiency analysis of conditions in outcome A_1

Conditions	Necessity	Sufficiency
a_1	0.984573	0.669356
$\sim a_2$	0.798814	0.843841
a_2	0.977057	0.795491
a_3	0.901899	0.835624
a_4	0.935918	0.715020
$\sim a_4$	0.742490	0.857371
a_5	0.907437	0.789809

Assumptions a_1 (present/ absent).
a_2 (present).
a_3 (present).
a_4 (present).
a_5 (present).
Frequency cutoff: 1
Consistency cutoff: 0.915,578

This solution shows high consistency and coverage, and the pathways make sense according to both theory and case knowledge. The final pathways of the resilient community can then be seen within the following pathways (Table 6.7):

(a) Suitable emergency response to basic public needs (a_1) AND high community engagement (a_5) AND insufficient shelter and school ($\sim a_2$)
(b) Suitable emergency response to basic public needs (a_1) AND sufficient shelter and school (a_2) AND improved public capabilities (a_4) AND resilient housing design (a_3)
(c) Sufficient shelter and school (a_2) AND high community engagement (a_5) AND unimproved public capabilities ($\sim a_4$)

(2). The negations of the conditions considered in the resilient society have formed the necessity and sufficiency of each condition impacting the non-resilient community.

The model has considered both the absence or presence of a_1 while all the rest of the conditions were absent. Additionally, the authors have selected $a_1 * a_2 * \sim a_3$ as the prime implicant (Table 6.8).

Table 6.7 Success pathways of outcome A_1

Solutions	Raw coverage	Consistency
$a_1 * \sim a_2 * a_5$	0.798814	0.886987
$a_2 * \sim a_4 * a_5$	0.714822	0.9584
$a_1 * a_2 * a_3 * a_4$	0.847035	0.926903

Solution coverage: 0.93,419. Solution consistency: 0.852787

Table 6.8 Necessity and sufficiency analysis of conditions in outcome ~ A_1

Conditions	Necessity	Sufficiency
~a_1	0.586306	0.978108
a_1	0.994616	0.794944
a_2	0.874074	0.836071
~a_2	0.786339	0.975783
~a_3	0.849092	0.910518
~a_4	0.682705	0.926061
~a_5	0.794583	0.909844
a_5	0.896198	0.917025
a_3	0.841353	0.916438

Table 6.9 Failure pathway of outcome A_1

Solutions	Raw coverage	Consistency
$a_1 * a_2 * \sim a_3$	0.785522	0.93507

Solution coverage: 0.785522. Solution consistency: 0.93507

Assumptions a_1 (present/absent).

~a_2 (absent).

~a_3 (absent).

~a_4 (absent).

~a_5 (absent).

Frequency Cutoff: 1

Consistency Cutoff: 0.927582

The pathways for the non-resilient society are presented below (Table 6.9):

Sufficient emergency answer to basic public needs (a_1) AND adequate shelters and schools (a_2) AND unreasonable housing design (~a_3).

6.5 Success and Failure Pathways for Each KPI (Outcome)

This section provides a detailed evaluation of the research findings, recognizing the causal combinations of conditions that lead to the success or failure of outcomes and the main objective shown in Fig. 6.1. Conditions that did not show a strong theoretical link to the outcomes were considered to be present and absent simultaneously. The process started with calculating necessity and sufficiency scores of individual conditions. Note that the order of the conditions in the pathways does not imply the sequence of their occurrence. In each path, the conditions must be present or absent (shown by ~) to gain the outcome. The outcome can be achieved through any of the pathways.

6.5.1 *Achieving Resilient Society*

Although all the conditions in this outcome's success pathways (except $\sim a_2$ and $\sim a_4$) have shown a high necessity score, unimproved public capabilities to cope with disasters ($\sim a_4$) have gained the highest sufficiency (sufficiency = 0.86). To evaluate how different conditions interact to affect the outcome, the combinations of conditions (causal configurations) have been evaluated, which are shown in (Fig. 6.5). The three recognized success pathways for a resilient society appear in all the neighborhoods identified as successful and show overall consistency of 0.85 and coverage of 0.93.

Interestingly, the availability of shelters and schools (condition a_2) appeared in all the solutions; however, it was present in the second and third pathways but absent in the first one. Mahdab, Razmandegan town, Amir Kabir & Baghkhan, and Lorestaniha were identified as resilient neighbourhoods, although they have a shortage of safe shelters and schools during the transitional phase ($\sim a_2$). However, the rescue process in those neighbourhoods was highly satisfactory (condition a_1), as residents had adequate access to sanitation facilities, and the average start time of the rescue was shorter than in the other neighbourhoods. In addition, public engagement (condition a_5) was sufficient in those neighbourhoods, contributing to the success of outcome A_1.

Ansari and Fakhr Abad had access to adequate safe shelters and education after the earthquake (a_2). Ansari neighbourhood, however, has revealed low improvement in public capabilities to cope with future disasters ($\sim a_4$). Based on the respondents' answers, very few people have been trained. Therefore, the public is not well-informed on how to tackle future disasters. However, the households were highly engaged in the recovery project as the Ansari neighbourhood has the highest public participation score. Therefore, it is expected that although formal training is lacking, people's participation has caused informal knowledge transmission, improving public knowledge about safe construction (Tran, 2015).

On the other hand, the Fakhr Abad neighbourhood presented the second pathway, which revealed that a resilient society could be achieved if the community receives adequate assistance in the emergency, transitional, mid-term, and long-term stages of recovery. It is quite surprising that reasonable housing design is only presented in the second pathway. Although it gained high necessity, it did not reach the sufficiency level (necessity = 0.90 and sufficiency = 0.84). However, its absence is significantly sufficient to fail to achieve a resilient society, although not necessarily leading to failure (necessity = 0.85 and sufficiency = 0.91). The failure pathway of outcome A_1, presented in Fig. 6.6, was present in all unsuccessful neighbourhoods. Sufficient emergency response to basic public needs (condition a_1) is the only necessary condition for failure whose combination with sufficient shelters and schools (condition a_2), and unreasonable housing design ($\sim a_3$) resulted in failure in outcome A_1. This pathway has gained a score of 0.94 for consistency, while the coverage score is 0.79.

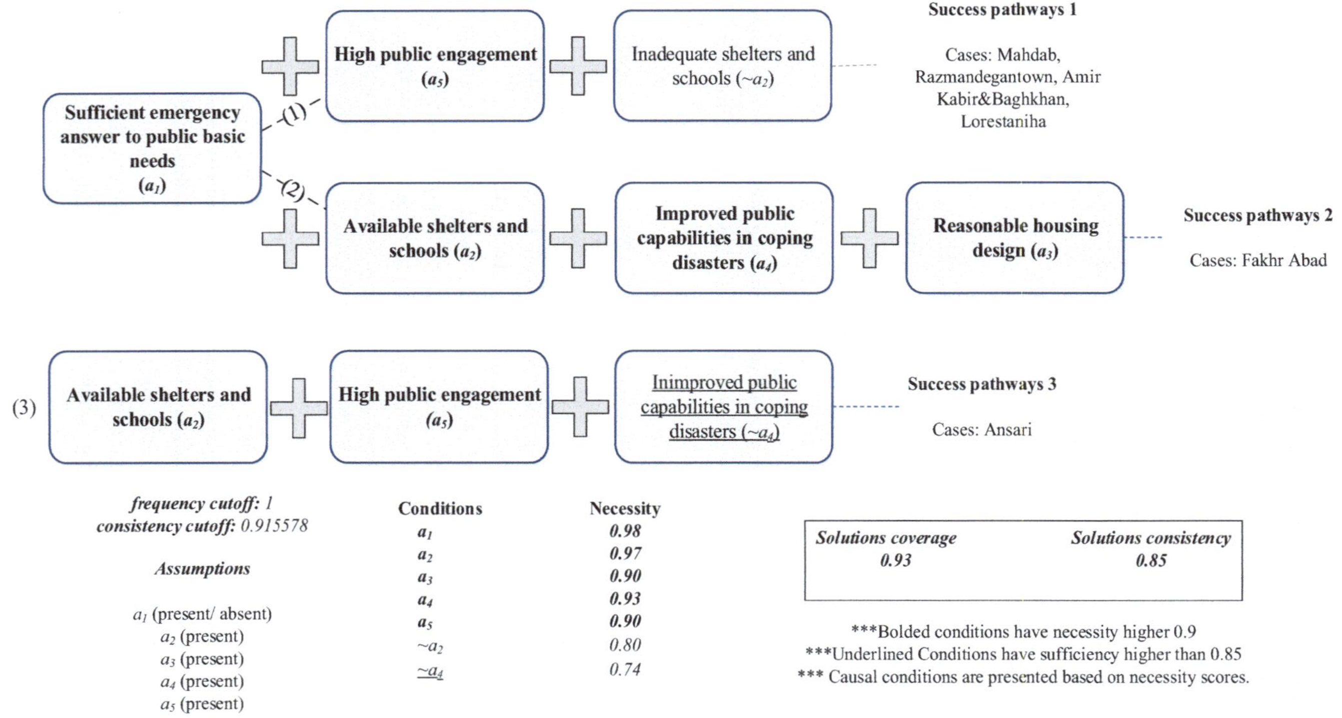

Fig. 6.5 Success pathways to outcome A_1

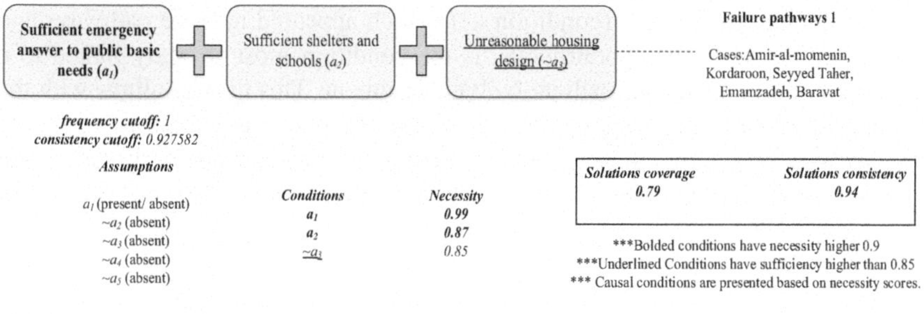

Fig. 6.6 Failure pathways to outcome A_1

6.5.2 Achieving Sustainable and Resilient Built Environment

The necessity and sufficiency analysis of individual conditions regarding outcome A_2 has shown that although all the single conditions, except the insufficient emergency answer to basic public needs (~a_1) and shortage of shelters and schools (~a_2), are necessary, only those exceptions have enough sufficiency (sufficiency of ~$a_1 = 0.96$, the sufficiency of ~$a_2 = 0.87$). However, the causal recipes show different combinations of the conditions that might lead to this outcome. Four pathways were identified in eight neighbourhoods to achieve a sustainable and resilient built environment. The solution's overall consistency and coverage scores are 0.84 and 0.99, respectively. The success and failure pathways are presented in Fig. 6.7.

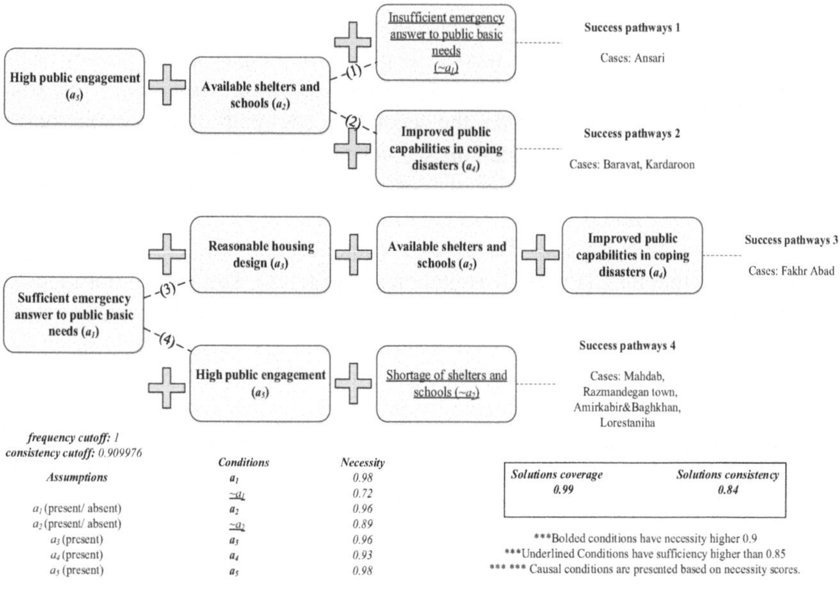

Fig. 6.7 Success pathways to outcome A_2

High public engagement (condition a_5), which appeared in three pathways and was present in seven neighbourhoods, is the condition most strongly linked to a successful, sustainable and resilient built environment. This finding aligns with the results of previous researchers, who stated that city planning should consider the real needs of people by engaging senior citizens and civic groups in disaster planning scenarios and organizing frequent neighbourhood events (Sadri et al., 2018). However, people's participation should not only be limited to planning and decision-making. It is necessary to keep people updated about the recovery process. The Ansari neighbourhood presented the first path and had the highest community participation score and a remarkable public information sharing score. Although informing people is at the third level on the ladder of community participation (Flores Salas, 2016), it can positively affect a sustainable and resilient built environment. Finally, people's participation in housing was a critical factor in Bam's neighbourhoods. Most households in the Ansari neighbourhood were in charge of housing, acting as managers of their houses' construction and working under the government's guidance. High community participation in the reconstruction phase can result in a sense of ownership and increased satisfaction (Xu et al., 2019).

According to the interviews and the damage zoning maps, the Razmandegan neighbourhood, located at the south corner of the city, did not have many residents at the time of the earthquake. Since it was further from the earthquake fault and experienced more minor damage. However, after the quake, nearby villagers who had experienced less damage from the earthquake rushed to this neighbourhood, causing insufficient available shelters. On the contrary, the Ansari neighbourhood near the city centre was ranked as a second-level destroyed neighbourhood, receiving an inadequate emergency response. However, the residents in both neighbourhoods positively addressed the city's changes after the earthquake. Therefore, one can conclude that city infrastructure can efficiently recover if people participate in all phases of the recovery process, even though the assistance received in the emergency OR transitional stages is not adequate. Like the second success path of a resilient society (outcome A1), Fakhr Abad presented the second pathway for the success of outcome A2, revealing that sufficiently managed assistance throughout the project could also lead to a sustainable and resilient built environment.

Amir-al-momenin, Seyyed Taher, and Emamzadeh were the unsuccessful neighbor-hoods that failed to achieve outcome A_2 (Fig. 6.8). These neighborhoods all presented one pathway. Sufficient emergency response to basic public needs (condition a_1) and sufficient shelters and schools (condition a_2) were the only necessary conditions observed in the failure pathway. At the same time, the latter condition also gained high sufficiency (sufficiency = 0.91). The presence of individual conditions such as unreasonable housing design ($\sim a_3$) and low public engagement ($\sim a_5$) guaranteed failure since their sufficiency was high (sufficiency of $\sim a_3 = 0.97$ and sufficiency of $\sim a_5 = 0.99$).

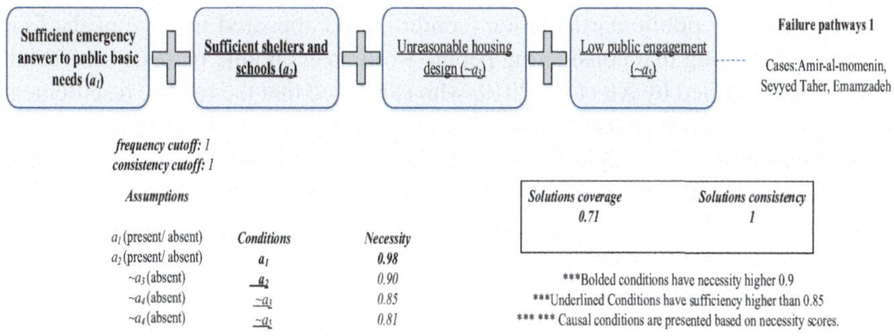

Fig. 6.8 Failure pathways to outcome A_2

6.5.3 *Achieving Resilient Economy*

The individual condition's sufficiency and necessity analysis revealed that all conditions, except insufficient emergency answers to basic public needs (~a_1) and shortage of shelters and schools (~a_2), are necessary to gain outcome A_3. However, ~a_1 and ~a_2 conditions are the only sufficient conditions (sufficiency of ~$a_1 = 0.96$, sufficiency of ~$a_2 = 0.85$). The study identified four possible causal configurations leading to a resilient economy; the overall consistency and coverage scores were 0.84 and 0.99. The pathways for success and failure of a resilient economy are illustrated in Fig. 6.9.

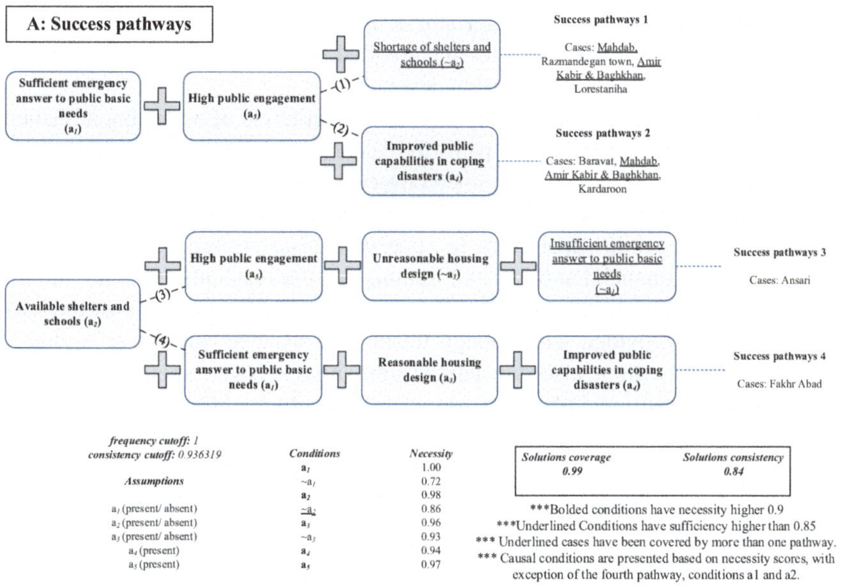

Fig. 6.9 Success pathways to outcome A_3

A high level of public participation (condition a_5) appeared in three of the four pathways, indicating that considering people's role in economic recovery is crucial. This fact was implied by Xu et al., 2019, who explained that the remote resettlement of households could increase livelihood vulnerability (Xu et al., 2019). A lack of considering the role of people in long-term recovery was an obstacle in gaining a successfully developed economy after the Wenchuan earthquake in China 2008. The survival of small businesses, especially disaster tourism, requires considering the people's role in long-term economic recovery (Li, 2017; Wang et al., 2013).

Looking at the pathways, emergency response to basic public needs (condition a_1), and availability of shelters and schools (a_2) appeared across the majority of the neighbourhoods and pathways; however, both their presence and absence were observed. Several studies have explored the importance of rescue operations and the availability of shelters concerning resuming livelihood (Jigyasu, 2013). A poor rescue operation is one of the bottlenecks in recovery projects following disasters (Xu & Lu, 2012). Evidence from actual cases and studies has emphasized short temporary accommodations during emergencies and transitional phases due to adverse effects on livelihood recovery (Di Gregorio & Soares, 2017).

However, this study's first and third causal configurations imply those conditions (conditions a_1 and a_2) might interact differently, leading to a resilient economy. While the households in the Amir Kabir & Baghkhan neighbourhood had significantly low access to shelters and education, and residents lived in the temporary shelters for many months, the rescue team arrived within one day after the earthquake. The residents had sufficient access to sanitation and medical facilities. On the contrary, Ansari's households faced difficulty getting access to sanitation facilities, and the rescue started much later than in the Amir Kabir & Bghakhan neighbourhood. Although the residents were more satisfied with the provided services in the transitional stage, they were less content with the housing design.

Kordaroon, the most successful neighbourhood in terms of achieving a resilient economy, is presented in the second pathway. According to the interviews, this neighbourhood experienced a high economic recovery due to a high employment rate and increased business after the earthquake. This neighbourhood showed one of the best improvements in public capabilities to cope with disasters (condition a_4), representing adequate financial support and training for the residents. This finding has been supported by a previous study in Bam that reported fast recovery of the businesses that were provided with enough financial assistance (Fallahi & Arzhangi, 2011). This neighborhood also showed relevantly high access to shelters and education during the transitional phase (condition a_2). Locating at the city centre and the city's old market, people were highly engaged in the recovery (condition a_5).

Moreover, the first and second success pathways emphasize high people participation (a_5) and sufficient emergency answers to basic public needs (a_1). While improvement in public capabilities tackling imminent disasters (a_4) must be present in the second path, the shortage of shelters and schools ($\sim a_2$) has appeared in the other pathway observed in four cases. The fourth success pathway is the common path to a resilient society (outcome A_1) and a sustainable and resilient built environment (A_2).

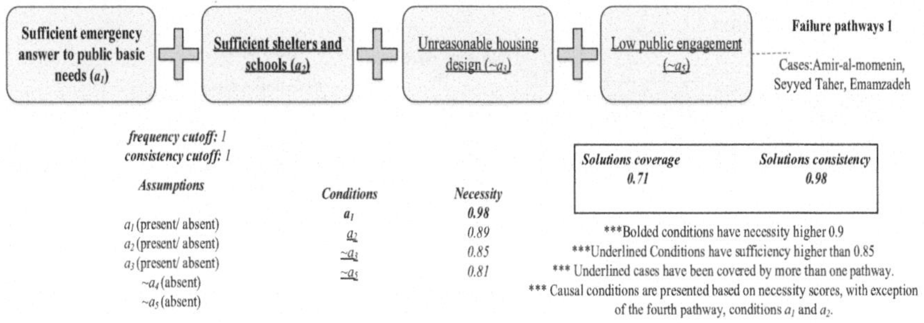

Fig. 6.10 Failure pathways to outcome A_3

The negation of outcome A_3 in Fig. 6.10, illustrating non-resilient recovery, presented one path-way with acceptable coverage and consistency scores. This solution appeared in all the unsuccessful neighbourhoods, demonstrating that these neighbourhoods that presented limited people participation ($\sim a_5$) and not climatically and culturally designed houses ($\sim a_3$) failed to reach a resilient economy, although their emergency and transitional needs were answered (conditions a_1 and a_2).

6.5.4 Achieving Comprehensive and Long-Lasting Success of SNDR Projects (Communities' Perspective)

The comprehensive, successful recovery project following the Bam community perspectives has been conducted by assigning the average score of the last calculated outcomes, giving twice the weight to A_3. There were no strong theoretical links between this outcome and public capabilities in coping with disasters (a_4) and public engagement (a_5), which are intangible results and may not be considered important by the public. Therefore, those conditions have been assumed to be both present and absent. The necessity and sufficiency analysis of the individual conditions showed that all the conditions, except unimproved public capabilities in coping with disasters ($\sim a_4$), are necessary. In contrast, only reasonable housing design (a_3) and unimproved public capabilities in dealing with disasters ($\sim a_4$) are sufficient for the ultimate outcome (sufficiency of $a_3 = 0.86$, the sufficiency of $\sim a_4 = 0.86$). The fsQCA, finally, has suggested three pathways leading to this outcome illustrated in Fig. 6.11. The solution represents all six successful neighbourhoods with a coverage score of one and a consistency score of 0.88.

Presented by three pathways and four cases, sufficient emergency response to basic public needs (condition a_1) is strongly linked to a successful, comprehensive and long-lasting recovery per the residents' perspectives. The high necessity score of this condition indicated that rescue operations should be promptly conducted to reach this outcome. In addition, high public engagement (condition a_5), which was observed in three causal configurations and five neighbourhoods, was consistent with

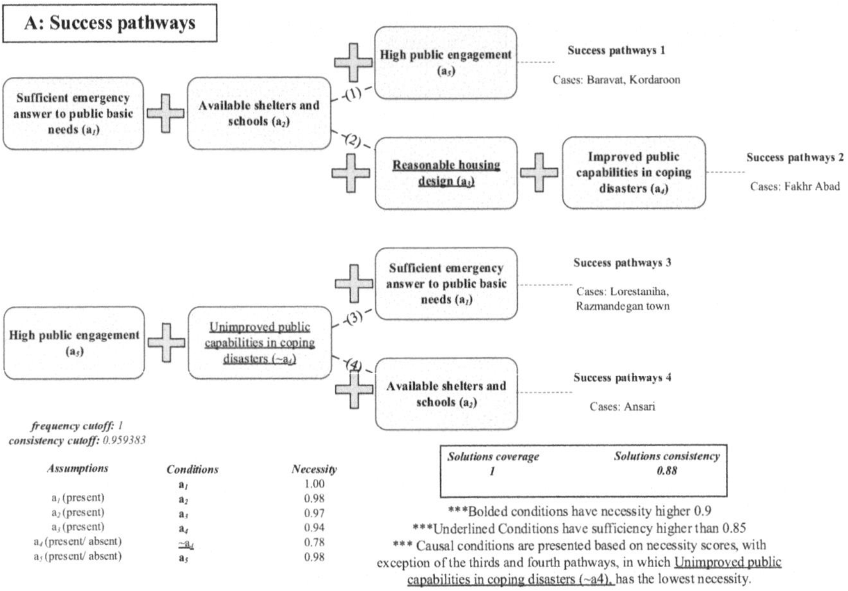

Fig. 6.11 Success pathways to the ultimate outcome

previous studies concerning the role of public engagement in the overall success of a recovery project (Dhamangaonkar, 2019; Sadiqi et al., 2017). Neighbourhoods such as Ansari were highly engaged in the planning, design, and construction phases. Therefore, the Ansari neighbourhood experienced a successful, comprehensive and long-lasting recovery from the residents' perspectives.

The simultaneous presence of sufficient emergency response to basic public needs (condition a_1) and high public engagement (condition a_5) was observed in the first and third pathways. While Baravat and Kordaroon residents had adequate access to safe shelters and education after the earthquake (condition a_2), Lorestaniha and Razmandegan neighbourhoods demonstrated low improvement in public capabilities to respond to a disaster ($\sim a_4$) since most residents were not provided with disaster training. However, all four neighbourhoods have successfully achieved the ultimate outcome, indicating the necessity of considering the different combinations of the mentioned conditions.

Two of the pathways were separately presented by Fakhr Abad and Ansari neighbourhoods; both neighbourhoods included sufficient shelters and schools in the transitional phase (condition a_2). The latter condition showed a high necessity score and appeared in three pathways and four neighbourhoods. However, the Fakhr Abad neighbourhood established a relatively more remarkable improvement in public capabilities to respond to disasters (condition a_4) than the Ansari neighbourhood. The majority of the residents from the Fakhr Abad neighbourhood participated in disaster training, which was mostly conducted before the earthquake. Moreover, most households received financial assistance lasting one year after the earthquake. On the contrary, the Ansari residents received less disaster training and although they had access to financial aid, it did not last long enough. However, both communities achieved successful recovery from the residents' perspectives. Therefore, it is expected that, although formal training was lacking in the Ansari neighbourhood, public engagement resulted in informal knowledge transmission, which improved the public's knowledge about safe construction (Arefian Farnaz, 2018).

Additionally, the Fakhr Abad neighbourhood presented a necessary and sufficient condition. Reasonable housing design (a_3), which was only observed in Fakhr Abad, was essential to reaching this outcome. In addition, climatically and culturally built houses can guarantee to achieve success. The findings in the literature also supported this observation. As Yilmaz et al., 2013 explain, the government's efforts to plan standardized houses for all damaged regions may negatively impact house designs and structures (Yilmaz et al., 2013). After the Wenchuan earthquake, designers' unfamiliarity with the cultural and local features of the damaged area resulted in inappropriate designs and dissatisfaction among residents (Xiao et al., 2020).

A close look at the part of the study's truth table (Table 6.10) demonstrates that a minor difference in the causal configuration leads to a major change in the results. These Insufficient but Necessary components of Unnecessary but Sufficient causes are called INUS causes (Befani, 2016). Considering unique circumstances, the presence or absence of one condition can make a significant difference in the pathways. A comparison between the following casual configurations revealed that people's participation is critical, and absence results in failure. While the neighbourhoods showed similarity in the combination of the conditions, except for condition a_5, the Kordaroon and Baravat neighbourhoods successfully reached the ultimate outcome. In contrast, low public engagement (condition a_5) in the Emamzadeh, Amir-al-momenin, and Seyyed Taher neighbourhoods resulted in failure to achieve the ultimate outcome.

The solution for non-successful recovery, showing two possible pathways covering all five unsuccessful neighbourhoods, is illustrated in Fig. 6.12. The solution's coverage is 0.8, and its coverage is perfect. Improved public capabilities in coping with disasters (condition a_4) is necessary for the outcome's failure, although it is not sufficient. Unreasonable housing design ($\sim a_3$), low public engagement ($\sim a_5$), and shortage of shelters and schools ($\sim a_2$) are highly sufficient, although their necessity score is lower than 0.9 (sufficiency of $\sim a_3 = 0.97$, sufficiency of $\sim a_5 = 0.98$ and, sufficiency of $\sim a_2 = 0.98$).

The failure pathways presented in Fig. 6.12 both share improved public capabilities in coping with disasters (condition a_4) and unreasonable housing design ($\sim a_3$).

Table 6.10 INUS cause of the ultimate outcome

Neighbourhood name/condition	a_1	a_2	a_3	a_4	a_5	Ultimate outcome
Kordaroon, Baravat	1	1	0	1	1	1
Emamzadeh, Amir-al-momenin, Seyyed Taher	1	1	0	1	0	0

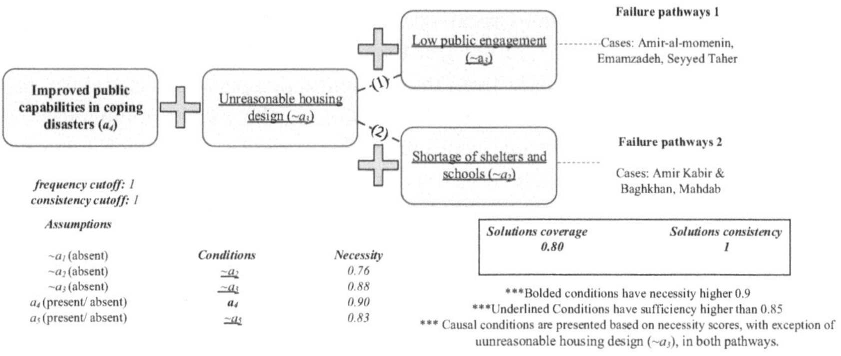

Fig. 6.12 Failure pathways to the ultimate outcome

Amir-al momenin, Seyyed Taher, and Emamzadeh can be observed in the first failure pathway and the second solution presented Amir Kabir & Baghkhan and Mahdab. Seyyed Taher presented the lowest score for reasonable housing design (~a_3), and the public was not actively engaged in that neighbourhood. More than half the respondents in Seyyed Taher preferred living in their old houses, and all of them had made modifications to their housing design. Moreover, public engagement in all recovery phases was significantly low (~a_5). In addition, the children did not have access to education during the transitional phase in the Mahdab neighbourhood, which had the lowest score in shelter and education availability (~a_2). The residents also lived in transitional houses for years after the earthquake. The combination of the mentioned shortcomings has resulted in those neighbourhoods failing to achieve successful recovery per the residents' perspectives; however, low public engagement may have had a more significant impact on the failure due to its higher necessity score.

6.6 Spearman Correlation Coefficient Analysis of the CSFs

Although the presented KPIs and CSFs have been applied in the Bam recovery project and the success/failure pathways (combinations of CSFs lead to KPIs) have been identified, the correlations among the CSFs need further exploration. Due to the not normally distributed sample, including nominal and ordinal data, a non-parametric correlation coefficient analysis method, Spearman's rank correlation, was applied in this study to measure the statistical dependence between two variables. Except for "water availability in the shelters (*b1-1*)" and "education availability (*b3*)", a correlation coefficient test was applied for all the conditions' factors in Table 6.2. Spearman's rank correlation coefficients (r) and p-values were determined between all the factors' indicators. Correlation coefficients $r > 0.400$ or < -0.400 were considered as strong correlations, while $r < 0.20$ or > -0.200 showed weak correlations, and middle-ranking showed moderate correlations (Akoglu, 2018). Spearman correlation coefficient was calculated by Eq. 6.5. Additionally, two-tailed *p*-value tests were conducted, and statistical significance was determined at $\alpha = 0.05$ or $\alpha = 0.001$ to accept or reject the null hypothesis shown below. The model has been analyzed using SPSS 26.

H0 = The correlation coefficient isn't essentially diverse from 0. (No direct correlation between the specified variables.)

H1 = Factors' correlation coefficient is essentially distinctive from 0. (There's a direct correlation between the specified variables.)

$$r_s = 1 - \frac{6 \sum d_i^2}{n(n^2 - 1)} \tag{6.5}$$

where,

$\boldsymbol{r_s}$ = Spearman correlation coefficient.

$\boldsymbol{d_i}$ = difference within the ranks of two factor values for each item

$\boldsymbol{n}$ = total number of observations.

Table 6.11 illustrates the significant correlations; however, as there is no theoretical evidence showing the correlations between "waiting time to find emergency shelter (*b4-1*) and living experience, preference of old or new house (*b5-1*)" and "waiting time to find emergency shelter (*b4-1*) and living experience, preference of old or new house (*b5-1*) and house climate design (*b6*)", those findings have not been presented in the conditions' correlation model.

The main findings are listed below, and the correlations are graphically demonstrated in Fig. 6.13:

(1). Correlation analysis results illustrate that the recovery time has negatively affected the public access to recovery information. According to Table 6.11, There was a moderate positive correlation between "informing public about recovery process" and "time household waited for rescuers" ($r = 0.248, p = 0.01$) and between "informing public about recovery process" and "time household waited for having shelter" ($r = 0.207, p = 0.05$.). However, the transitional housing stage negatively correlated with "informing the public about the recovery process" ($r = -0.195, p = 0.05$). Moreover, community training, especially before the earthquake, has efficiently enhanced public recovery information" ($r = 0.201, p = 0.05$). It is also observed that well-informed people had more access to financial assistance ($r = 0.211, p = 0.05$).

(2). Medical service access is slightly more correlated to the emergency phase ($r = 0.249, p = 0.01$) than the transitional recovery stage ($r = 0.203, p = 0.05$). In the aftermath of a disaster, when the number of injured is higher, there is more need for medical treatment; however, as time passes, the recovery procedure deals with the emergency needs and goes forward to answer other needs. However, access to medical services is negatively correlated to livelihood development, which indicates the weakness in balancing the distribution of financial and medical aids ($r = $ -–0.226, $p = 0.01$).

Table 6.11 CSFs' correlation coefficient results

b1-2	b1-2	b2-1	b4-1	b4-2	b4-3	b5-1	b5-2	b6	b7	b8	b9-1	b9-2	b10
b1-2	1.000												
b2-1	0.063	1.000											
b4-1	0.160	0.415**	1.000										
b4-2	0.249**	0.115	0.034	1.000									
b4-3	0.203*	−0.051	−0.110	0.342**	1.000								
b5-1	0.090	0.137	0.258**	0.047	−0.109	1.000							
b5-2	0.088	−0.084	−0.128	−0.055	0.081	0.090	1.000						
b6	−0.090	0.126	0.179*	0.021	−0.160	−0.105	−0.286**	1.000					
b7	0.072	0.190*	0.134	0.166	0.084	0.105	0.000	0.085	1.000				
b8	−0.226**	0.004	0.003	−0.024	−0.060	0.054	−0.162	−0.024	0.092	1.000			
b9-1	−0.032	0.068	0.115	−0.058	−0.004	−0.015	−0.028	0.168*	0.032	−0.094	1.000		
b9-2	0.013	0.248**	0.207*	0.045	−0.195*	0.211*	−0.080	0.161	0.201*	0.211*	0.006	1.000	
b10	0.007	−0.082	0.025	−0.034	−0.076	0.018	−0.084	0.103	−0.071	−0.069	−0.095	−0.113	1.000

* Correlation is significant at the 0.05 level (2-tailed)
** Correlation is significant at the 0.01 level (2-tailed)

(3). There was a weak positive correlation between "participation in site selection" and "house climate design" ($r = 0.168, p = 0.05$). Additionally, considering the moderate negative correlation between "house climate design" and "modification in new house layout", ($r = -0.286, p = 0.01$), the cyclic correlations between those three conditions are visible. Appropriate site selection may positively affect the climate of housing design and vice versa. In the case of unsuitable climatic houses, the residents tended to modify the house layout.

(4). The strongest correlation has been seen between "time household waited for having shelter" and "time household waited for rescuers" ($r = 0.415, p = 0.01$), which is logical. The faster the rescue starts, the sooner the survivors can access shelters.

(5). The Spearman test has identified a moderate positive correlation between "transitional housing living time" and "time household lived in shelter" ($r = 0.342$, $p = 0.01$). The Bam recovery project revealed that the longer survivors live in the shelters, the longer they have lived in the transitional houses.

(6). Interestingly, public participation in the reconstruction stage did not show any correlation with the other conditions. However, as the results from the last sections (fsQCA) indicate, combinations of public participation with other conditions can significantly affect the recovery process. In contrast, in some cases, the absence of public engagement has resulted in failure to reach the outcomes.

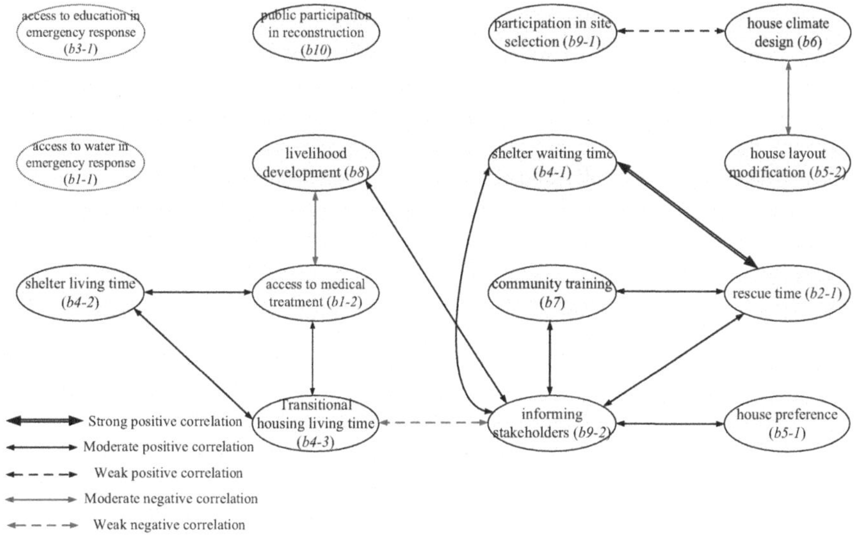

Fig. 6.13 Correlations between CSFs/ their indicators

6.7 Chapter Summary

This chapter demonstrates how communities can successfully recover after such a severe earthquake. Although the disaster recovery evaluation was mostly conducted considering separate factors, this study attempted to evaluate how combinations of the elements might lead to the success or failure of a recovery project. Moreover, correlations among the CSFs have been assessed. The original contributions of this chapter are outlined below.

- The research findings illustrate the determinant role of public engagement and emergency response in the recovery process. This finding emphasizes the need for active engagement of all people in all phases of recovery. However, public participation in the reconstruction stage did not correlate with the other CSFs' indicators, indicating that although its combination with others significantly impacted the KPIs, it does not correlate with the rest of the CSFs.
- A combination of sufficient emergency response (rescue and health service), adequate response through to the transitional phase (shelter availability and living duration in permanent houses and shelters), climatically and culturally designed houses, and enhanced public strength through financial aid and disaster training could achieve recovery success. The Fakhr Abad neighbourhood, presenting the mentioned pathway, performed full economic, social, and built environment recovery and ranked as a successful project per the residents' perspectives. Therefore, it is necessary to provide sufficient answers to all the recovery phases, from rescue operations to public capability enhancement.
- One of the most interesting findings of this study is the significant role of housing design. Cultural housing design has been recognized as a crucial parameter affecting the satisfaction of residents (Tauber, 2015). The results of this study reveal that reasonable housing design is necessary to achieve all four outcomes. Moreover, the climatically and culturally designed houses guaranteed success in recovery based on Bam's neighbourhoods. On the other hand, the recovery project failed to reach all the outcomes in neighbourhoods with inappropriately designed houses.

Since the Bam earthquake was located in a historical city with significant characteristics as the case study selected for this research, it was not expected to obtain the same results from different disaster recovery projects. Further study is needed to identify the success pathways for various disasters in other countries. Moreover, as the Bam earthquake happened more than one decade ago and the research used primary data collection by conducting public interviews, the data was refined based on the researchers' observations. Finally, since SNDR projects involve many stakeholders, it is crucial to structure further studies to analyze the recovery pathways following various beneficiaries' points of view.

References

Akoglu, H. (2018). User's guide to correlation coefficients. *Turkish Journal of Emergency Medicine, 18*(3), 91–93. https://doi.org/10.1016/j.tjem.2018.08.001

Arefian Farnaz, F. (2018). *Organising post-disaster reconstruction processes process. Housing reconstruction after the Bam earthquake*. Springer. https://doi.org/10.1007/978-3-319-70911-6

Befani, B. (2016). *Pathways to Change: Evaluating development interventions with Qualitative Comparative Analysis (QCA)*. Elanders Sverige AB.

Dhamangaonkar, V. (2019). *Identification of legal and policy barriers to timely post-disaster reconstruction and strategy development for effective recovery processes in United States*. ProQuest Dissertations and Theses, The University of Texas. https://doi.org/.1037//0033-2909.I26.1.78

Di Gregorio, L. T., & Soares, C. A. P. (2017). Post-disaster housing recovery guidelines for development countries based on experiences in the American continent. *International Journal of Disaster Risk Reduction, 24*(2017), 340–347. https://doi.org/10.1016/j.ijdrr.2017.06.027

Fallahi, A., & Arzhangi, S. (2011). Business continuity after the 2003 Bam Earthquake in Iran. *Communications in Computer and Information Science, 257 CCIS*, 532–540. https://doi.org/10.1007/978-3-642-27207-3_58

Flores Salas, A. (2016). *Evaluation of the use of lightweight concrete panels for post disaster house reconstruction using building information modelling*. University of Manchester.

Jigyasu, R. (2013). Long-term cultural impacts of disaster decision-making: The case of post earthquake reconstruction in Marathwada, India. *International Journal of Architectural Research, 7*(3), 14–23. https://doi.org/10.26687/archnet-ijar.v7i3.46

Jordan, E. (2012). *Pathways to community recovery: A qualitative comparative analysis* of post-disaster outcomes. University of Colorado.

Li, Q. (2017). *The reconstruction of the Sichuan earthquake: Fieldwork observation and examination of the news coverage of the Qiang ethnic group*. University of Delaware.

Ragin, C. C. (2008). *Redesigning social inquiry; fuzzy sets and beyond*. University of Chicago Press.

Ragin, C. C. (2009). Qualitative comparative analysis using fuzzy sets (fsQCA). In B. Rihoux & C. C. Ragin (Eds.), *Configurational comparative methods: Qualitative Comparative Analysis (QCA) and related techniques* (pp. 87–122). SAGE Publications, Inc. https://doi.org/10.4135/9781452226569.n5

Sadiqi, Z., Trigunarsyah, B., & Coffey, V. (2017). A framework for community participation in post-disaster housing reconstruction projects: A case of Afghanistan. *International Journal of Project Management, 35*(5), 900–912. https://doi.org/10.1016/j.ijproman.2016.11.008

Sadri, A. M., Ukkusuri, S. V., Lee, S., Clawson, R., Aldrich, D., Nelson, M. S., Seipel, J., & Kelly, D. (2018). The role of social capital, personal networks, and emergency responders in post-disaster recovery and resilience: A study of rural communities in Indiana. *Natural Hazards, 90*(3), 1377–1406. https://doi.org/10.1007/s11069-017-3103-0

Tauber, G. (2015). Architects and rural post-disaster housing: Lessons from south india. *International Journal of Disaster Resilience in the Built Environment, 6*(2), 206–224. https://doi.org/10.1108/IJDRBE-07-2013-0025

Tran, T. A. (2015). Post-disaster housing reconstruction as a significant opportunity to building disaster resilience: a case in Vietnam. *Natural Hazards, 79*(1), 61–79). https://doi.org/10.1007/s11069-015-1826-3

Wang, G. Z., Li, Y. M., & Liu, L. P. (2013). Relief effort and reconstruction after Wenchuan Ms8.0 earthquake. *Applied Mechanics and Materials, 275–277*, 2727–2734. https://doi.org/10.4028/www.scientific.net/AMM.275-277.2727

Xiao, Y., Olshansky, R., Zhang, Y., Johnson, L. A., & Song, Y. (2020). Financing rapid community reconstruction after catastrophic disaster: Lessons from the 2008 Wenchuan earthquake in China. *Natural Hazards, 104*(s1), 5–30. https://doi.org/10.1007/s11069-019-03789-9

Xu, H., Wang, L., Li, Y., Huang, L., & Yin, L. (2019). Critical Chinese government behavior for swift housing reconstruction programme - The case of Dujiangyan following Wenchuan Earthquake.

International Journal of Disaster Risk Reduction, 42, Article 101307. https://doi.org/10.1016/j.ijdrr.2019.101307

Xu, J., & Lu, Y. (2012). Meta-synthesis pattern of post-disaster recovery and reconstruction: Based on actual investigation on 2008 Wenchuan earthquake. *Natural Hazards, 60*(2), 199–222. https://doi.org/10.1007/s11069-011-0003-6

Yilmaz, D. G., Meding, J. V., & Erk, G. K. (2013). A theoretical approach to the design of a survey instrument in post-disaster reconstruction: Defining indicators for a human-based study in rural built-environment. *International Journal of Architectural Research, 7*(3), 40–56.

Chapter 7
From Policy Implementation to Public Perception: A Dual Approach to Assessing Disaster Recovery Success in the Wenchuan and Bam Earthquake Recovery Projects

Abstract The present approaches for disaster recovery assessment most often solely concentrate on the tangible outcomes officially documented by local authorities. Therefore, they face shortcomings in addressing impacted populations' complex experiences and perspectives. Top-down disaster recovery evaluation methods emphasize macro-level variables, including infrastructure restoration and economic revitalization. Conversely, bottom-up evaluation methodologies emphasize community-level perceptions, including recovery assessment, residents' emotional well-being, and satisfaction. By integrating top-down and bottom-up disaster recovery assessment methods, this study seeks to close this knowledge gap and provide a more comprehensive understanding of the disaster recovery assessment approaches that followed the 2008 Wenchuan and 2003 Bam earthquakes in China and Iran. This chapter found that while Wenchuan's recovery benefited from a more structured and comprehensive approach, insufficient public involvement presented significant challenges. In contrast, Bam's reliance on community participation underscored the significance of meticulous planning and accountability in attaining long-term recovery success. Temporal recovery trends over reconstruction years in Wenchuan revealed that recovery trends varied throughout the region. The local government's main priority was rehabilitating the built environment and infrastructure. However, the public regarded social and economic recovery as more important factors for a successful recovery after a disaster. Thus, it is imperative to prioritize comprehensive planning to balance long-term social and economic development strategies and immediate infrastructure needs.

Keywords Recovery evaluation metrics · Top-down recovery assessment · Bottom-up recovery assessment · Wenchuan earthquake

H. Bahmani et al., *Socio-Natural Disaster Recovery Projects*, Resilient and Friendly Living Environments, https://doi.org/10.1007/978-981-95-0113-7_7

7.1 Top-Down and Bottom-Up Disaster Recovery Assessment

Disaster recovery success is often assessed by considering tangible outcomes reported by local authorities. Most common measures focus on reconstructed facilities' quantity and speed of bounce back to normalcy, known as the macro-level or top-down recovery assessment method (Guo et al., 2024; Poudel & Shaw, 2024). This approach focuses on observable, concrete objectives like infrastructure rebuilding and economic recovery. Nevertheless, it frequently overlooks intangible factors such as people's well-being, emotional rejuvenation, and their relationship with newly constructed amenities (Fallahi & Arzhangi, 2011; Sichuan Provincial Department of Housing & Construction, 2013). From this perspective, top-down disaster recovery evaluation methods highlight quantifiable results. They emphasize the 'hard' facets of recovery by utilizing data provided by authorities, concentrating on evaluating progress through concrete indicators (Bahmani & Zhang, 2022a).

Since the top-down recovery assessment may not fully capture the nuanced on-the-ground experience and perception of the local residents affected by the disaster, less tangible aspects could be left aside. Bottom-up disaster recovery evaluation methods measure the success of recovery by collecting information from the people who were most directly impacted. These emphasize communal encounters, contentment with recovery initiatives, and a sense of reinstated normalcy. These assessments gather input on the recovery process from the community. They are crucial tools for comprehending the complex on-the-ground experience sometimes missed in top-down evaluations. Bottom-up recovery assessment approaches were commonly employed in previous research, such as the 2008 Wenchuan earthquake in China (Di et al., 2020; Zhang et al., 2018), the 2003 Bam earthquake in Iran (Bahmani & Zhang, 2022b; Omidvar et al., 2016), the 1999 earthquakes in Turkey (Johnson, 2006) and the 2001 Gujarat and 1999 Marmara earthquakes (Ganapati & Mukherji, 2014).

Both disaster recovery evaluation methodologies provide valuable insights but are seldom integrated. This division may hinder the capacity to assess recovery success and formulate effective strategic recommendations thoroughly. This study addresses this imbalance by incorporating top-down and bottom-up assessments, providing a more comprehensive understanding of recovery pathways and informing future recovery strategies. Note that in the disaster recovery studies, there is another terminology, "disaster recovery approaches", which is significantly different from what we seek in this study. "Disaster recovery approaches" refer to the operational strategies and processes used to restore normalcy following a disaster, whereas "disaster recovery assessment approaches" focus on evaluating these efforts. The latter requires developing evaluation indicators, collecting data, and applying analytical techniques to determine the effectiveness of recovery strategies.

On the other hand, the loop-planning idea of using the previous experiences gained from past disaster recovery projects has been proposed by several researchers (Griffey, 2010; Kermanshachi et al., 2019). Okada et al. 2018, evaluated community resilience and vulnerability by comparing disaster recovery projects in Japan and Australia. The study illustrated that the success of Japan's experience was due to applying *Macchizukuri* (connecting community members) in community recovery. Similar ideas have been examined through China and Japan's experience coping with the recent natural hazards to access civil society development in those countries (Akiyama, 2019). The broader perspective in cross-comparison has been applied in some other studies (Di Gregorio & Soares, 2017; Lu & Xu, 2015). The experience gained by American countries managing disaster recovery projects has been put forward as practical suggestions for developing countries (Di Gregorio & Soares, 2017). In a comprehensive research by (Lu & Xu, 2015), four major earthquakes in China, Iran, Japan, and the USA were appraised, and the lessons learned for earthquake management recovery planning were proposed. Although extensive research has been done comparing disaster recovery cases with various characteristics, disaster recovery projects after disasters are complex and dynamic projects entangling multiple beneficiaries (Guarnacci, 2012; Opdyke, 2017). Past studies have focused less on comparing disaster recovery assessment approaches, highlighting the need for further research on recovery evaluation methodologies in diverse conditions to inform balanced policies.

Present disaster recovery assessment methodologies often depend on top-down evaluation strategies that emphasize measurable results, often overlooking the actual experiences of the impacted population. Consequently, a comprehensive disaster recovery evaluation framework is required to assess disaster recovery outcomes from public and administrative viewpoints simultaneously. This study selected disaster recovery efforts following two major earthquakes in Iran and China as case studies. These countries illustrate considerable susceptibility to natural hazards and their unique vulnerabilities and coping strategies. The disaster recovery initiatives in Iran, especially after the Bam earthquake, are acknowledged for their significant public involvement. China's recovery strategy after the Wenchuan earthquake is characterized by its vertical, centralized administration. These distinctions make them suitable for investigating various recovery evaluation tactics and their efficacy in answering the stakeholders' needs. Moreover, based on the Iran and China risk index profile available by *INFORM Risk Index*[1], both countries are highly exposed to socio-natural hazards; on a scale of 0 to 10, China scores 7.5, and Iran scores 6.7. On the other hand, their vulnerability and coping abilities are very different. Iran has a lower vulnerability index of 1.3%, compared to China's vulnerability index of 4.3%., yet their scores for inadequate coping capacity are closer (4.4 for Iran and 3.5 for China). Despite similar levels of hazard exposure, these variations in vulnerability and coping mechanisms offer a compelling foundation for researching approaches to disaster recovery, resilience, and preparedness in each country, especially in the wake of large-scale earthquakes.

The Bam earthquake in Iran (2003) and the Wenchuan earthquake in China (2008) were among the most devastating seismic events in their respective regions in the early twenty-first century. In both cases, widespread destruction and significant loss of life occurred. Both earthquakes triggered varied responses from administrative

[1] Data obtained from: https://drmkc.jrc.ec.europa.eu/inform-index/INFORM-Risk.

authorities and local communities, making them ideal cases for analyzing disaster recovery assessment from both top-down (governmental) and bottom-up (community) perspectives. By examining these two events, this study aims to compare the effectiveness and challenges of each recovery evaluation approach. Therefore, this study chose the Bam earthquake in Iran (2003) and the Wenchuan earthquake in China (2008). To provide geographical context, Fig. 7.1 shows the locations of the Wenchuan earthquake in China and the Bam earthquake in Iran.

On December 26, 2003, a 6.7 magnitude earthquake hit Bam, affecting 142,000 people, among whom approximately 30,000 were injured, 100,000 were left homeless, and more than 26,000 lost their lives (Lu & Xu, 2015). The quake destroyed 85 per cent of that district's homes, city infrastructures, and many commercial centres (Arefian Farnaz, 2018). As the epicentre depth is located 10 km from the earth's surface, the most severely affected areas are located 15 km away, resulting in less

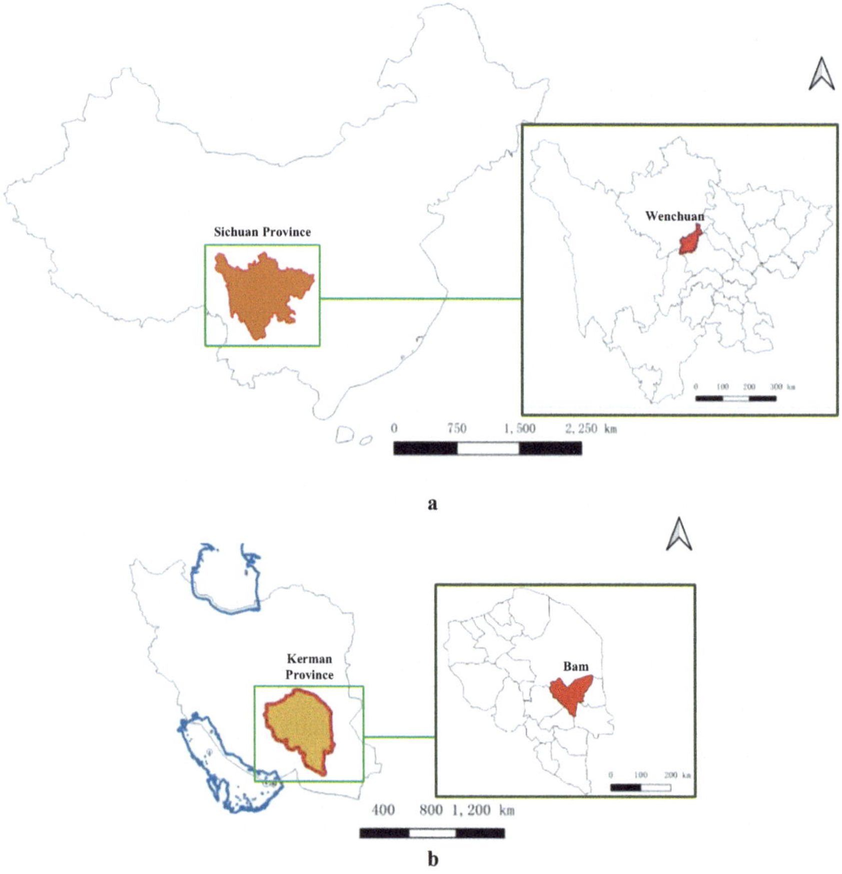

Fig. 7.1 **a** The location of the Wenchuan earthquake, **b** The location of the Bam earthquake

damage in nearby villages (Fallahi, 2003). A few years later, the Wenchuan earthquake occurred on May 12, 2008, with a magnitude of 8.0, in Sichuan Province, located in western China, a region less economically developed than the eastern part of the country. The ethnic and impoverished areas make this part of the country more vulnerable to disasters. This region is highly industrial, with many machinery manufacturers, metallurgy, chemical industries, and pharmacies (Lingling et al., 2008). The earthquake affected over half of China and Southeast Asian countries (Sichuan Provincial Department of Housing & Construction, 2013). Since the quake occurred at noon, when many children attended school and adults attended their jobs, high human loss occurred (Chen & Booth, 2011). As of July 24 (12 days after the quake), 69,197 people had been killed, 374,176 people injured, and 189 people missing. As of September 4, 2008, the direct economic loss was RMB 845.1 billion. In particular, the disaster losses in Sichuan Province owed more than 90 per cent of the total losses, nearly 80 per cent of Sichuan Province's total GDP in 2007[2] (Huang et al., 2011; Xiao et al., 2020). Based on available studies, the final death toll rate was recorded more than 80,000 (Lu & Xu, 2015).

This study seeks to combine top-down and bottom-up disaster recovery assessment methodologies to tackle the complexities in post-disaster recovery initiatives and improve subsequent disaster recovery policies. Therefore, this research benefited from the local authorities' reports (top-down approach) and public surveys (bottom-up assessment) as primary data sources. For top-down disaster recovery assessment, the study collected disaster recovery information from both cases by reading numerous official reports and published materials in English, Farsi, and Chinese. Regarding the bottom-up recovery assessment, the research team also conducted a public survey in the selected towns of Sichuan province to realize the public perception of post-disaster recovery. The results of this survey were subsequently compared to those from a comparable public survey performed during the Bam recovery effort, which had been previously released (Bahmani & Zhang, 2022b). Later, a top-down recovery assessment index was developed, incorporating key performance indicators derived from official reports and regional data. The recovery trajectory of the region's most severely suffered counties was systematically analyzed. This study highlights the need for a common understanding of disaster recovery procedures between governments and communities, advocating for comprehensive planning that balances short-term needs with long-term developments. By promoting government-community cooperation, the study provides actionable insights for enhancing the effectiveness and acceptance of recovery initiatives, which could be helpful for policy-makers, disaster management authorities, and local stakeholders. The rest of this paper is organized as follows: Sect. 7.2 presents the research materials and methodology by explaining the research framework, cases, and data collection methods. The study's detailed results are given in Sect. 7.3. The paper will delve

[2] This statistics were collected shortly after the earthquake, within the first few days, and continued to rise as more information became available. Due to the limited data openness, limited primary information, and the challenges in collecting disaster loss data, discrepancies in the reported loss statistics still exist.

into the discussion in Sect. 7.4. Conclusions, policy implementations, the study's limitations, and future research directions are given in Sects. 7.5 and 7.6, respectively.

7.2 Study Design

The study employed an integrated framework that merged two distinct recovery assessment techniques and utilised mixed methods to analyse its outcomes.

7.2.1 Research Framework

Current disaster recovery assessment methods tend to ignore the complicated experiences and viewpoints of the affected people in favour of officially recorded recovery outcomes. This study intends to bridge that gap by integrating top-down and bottom-up recovery evaluation methodologies. This framework would offer a comprehensive knowledge of disaster recovery assessment techniques by utilizing the 2008 Wenchuan and 2003 Bam earthquakes as case studies. This integrated structure allows us to evaluate disaster recovery properly and emphasizes the importance of combining several perspectives in disaster recovery assessment research to comprehensively and precisely assess recovery successes. By combining these methods, this framework guarantees that public and government viewpoints are considered when assessing the effectiveness of recovery measures. Figure 7.2 graphically walks the reader through this study's suggested research framework and methods.

7.2.2 Cross-Country Comparison of Recovery Assessment Approaches

- ***Top-down recovery assessment***: Cross-comparisoning different disasters is the most effective way to develop a disaster recovery management framework (Liu & Han, 2009). The study first put forward a detailed comparison of the recovery projects after the Bam and Wenchuan earthquakes, focusing on disaster recovery evaluation dimensions based on work published by Jordan & Javernick-Will, 2013 (Jordan & Javernick-Will, 2013). Through a cross-comparison of the recovery procedures after the two severe catastrophes selected for this study, readers will be provided with a detailed comparison of recovery projects focusing on social, economic, and built environment recovery from the government's perspective. The data were collected by reading numerous official reports, dissertations, books, and papers published in English, Farsi, and Chinese. Next, the Strengths, Weaknesses, Opportunities, and Threats analysis (SWOT) analysis has been applied for both

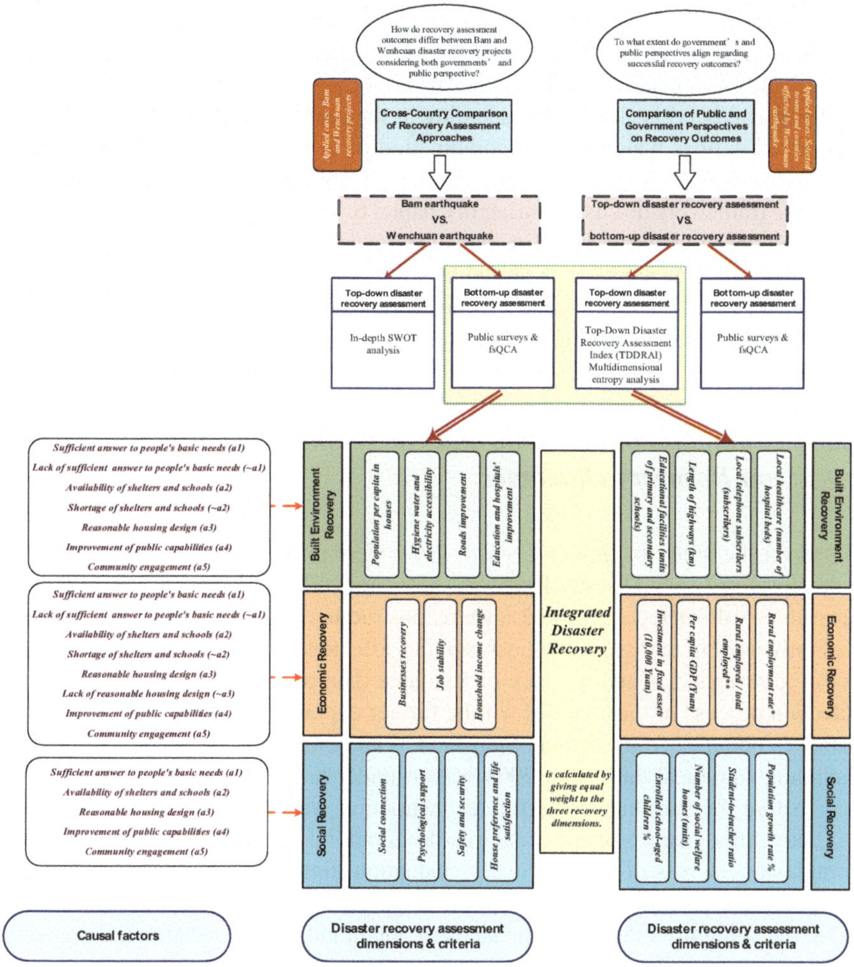

Fig. 7.2 Research framework and applied methods

cases, illustrating possible suggestions that can be helpful for policy-makers to see the ways for potential improvements in managing recovery projects.

- ***Bottom-up recovery assessment***: The bottom-up recovery evaluation was evaluated through public surveys and fuzzy-set Qualitative Comparative Analysis (fsQCA), based on the framework we previously applied to comprehend how Bam communities perceived recovery success pathways (Bahmani & Zhang, 2022b). The empirical data was collected by surveying selected villages in Sichuan Province (the case selection justification can be found in the following section). The data was later cleaned and prepared for applying fsQCA, which stands in the middle of qualitative and quantitative data analysis methods. This technique allows the researchers to examine several case studies, requiring an in-depth understanding of each case's context. At the same time, traditional statistical methods evaluate the dependent and independent variables within numerous cases, limiting the researchers' in-depth understanding of the cases (Befani, 2016). The fsQCA method defines which factors/ combination of factors might affect the result(s) (Devers et al., 2013). Three necessary steps in applying fsQCA were done based on the framework discussed earlier in Chapter 6. This chapter, however, revised the data calibration rubrics based on the field knowledge of its case studies. Therefore, the outcomes, an objective or result (desirable or undesirable) of the procedure, are three disaster recovery evaluation dimensions and integrated recovery (Fig. 7.1), obtained by equally averaging three recovery dimensions.

7.2.3 Comparison of Public and Government Perspectives on Recovery Outcomes

Based on the study by *Javernick-Will, 2013* (Jordan & Javernick-Will, 2013), an index for Top-Down Disaster Recovery Assessment (TDDRAI) was constructed to measure top-down recovery trends over reconstruction years. However, the TDDRAI was only used in ten selected cities/ counties affected by the Wenchuan earthquake due to discrepancies and the lack of adequate data on Bam recovery efforts (The selection process of the counties will be further explained in the next section). The data was obtained from Sichuan Statistical Yearbooks, 2008–2012[3] The reason for including statistical data from 2008 was to understand the region's condition the year before the earthquake. As the reconstruction was officially finished in 2011, we chose this year's statistical data as the temporal analysis endpoint. Next, the complexities across disaster recovery evaluation dimensions were analyzed using multidimensional entropy analysis. Assuming that m years, n cases, and α indicators exist, the study used the following steps to obtain indicators' weights and recovery scores.

$$X'_{ijk} = \frac{(X_{\mathrm{ijk}} - X_{kmin})}{(X_{kmax} - X_{kmin})}, \tag{7.1}$$

$$P_{ijk} = \frac{X'_{ijk}}{\sum_{i=1}^{m}\sum_{j=1}^{n} X'_{ijk}}, \tag{7.2}$$

$$E_k = \frac{-1}{\ln(n)}\sum_{i=1}^{m}\sum_{j=1}^{n} P_{ikj}\ln(P_{ikj}), \tag{7.3}$$

$$W_k = \frac{D_k}{\sum_{k=1}^{\alpha} D_k}, \tag{7.4}$$

[3] Each year's yearbook reports the previous year's statistical data for the region.

$$R_{ij} = \sum_{k=1}^{\alpha} W_k X'_{ijk} \tag{7.5}$$

where X_{ijk} is the value of the *k-th* indicator of case *j* in the year *i*, X'_{ijk} is the normalized value of X_{ijk}, $K_{\max}$ and K_{min} are maximum and minimum values of the indicator *k*, P_{ijk} is the weight of *k-th* indicator of case *j* in year *i*, E_k is Entropy value of *k-th* indicator, D_k and W_k are *k-th* indicator's the redundancy value and weight, and R_{ij} represents the total recovery score of case *j* in *i-th* year.

Next, we attempted to discover the similarities and differences between the public and government's points of view regarding recovery results after the Wenchuan earthquake by mathematically comparing the results obtained from top-down and bottom-up disaster recovery assessments.

7.2.4 Rationale Behind Case Studies Selection Affected by the Wenchuan Earthqauke

- **Bottom-up disaster recovery assessment; Public survey**

According to fsQCA principles, incorporating heterogeneity in case selection allows the analysis to capture various scenarios and conditions. This strategy could help identify different pathways to an outcome and understand the robustness of patterns across diverse contexts. In Wenchuan, to ensure the heterogeneity, we first considered death toll rates and chose three counties with the highest, middle, and lowest death numbers caused by the earthquake. Additionally, according to top-down recovery assessment results, we attempted to consider dissimilar recovery patterns across the cases. Wenchuan, therefore, with the highest human loss and lower top-down recovery scores, was selected. Mianzhu and Pengzhou, from middle and low-class levels of death toll rates, illustrated relatively higher top-down recovery scores over time. Next, we must ensure homogeneity in case selection and confirm that the cases have similar contexts and characteristics. Therefore, the study considered each county's two most severely affected towns, selected four villages (Fig. 7.3), and surveyed the 1st to the 7th of July 2024. Due to the challenges of effectively distributing online questionnaires and considering that much of the population was either low-educated or comprised of older residents, seven trained undergraduate students conducted face-to-face surveys with local residents. The participants must have resided in the region before the earthquake and were over 25 years old, resulting in 599 filled questionnaires. This approach could guarantee the homogeneity and heterogeneity of the cases to ensure that the analysis captures the relevant patterns and provides meaningful insights into the conditions leading to the outcome of interest.

- **Top-down disaster recovery assessment; Application of TDDRAI**

After the Wenchuan earthquake, due to the vast extent of the affected areas, the central government introduced the national counterpart assistance method, a

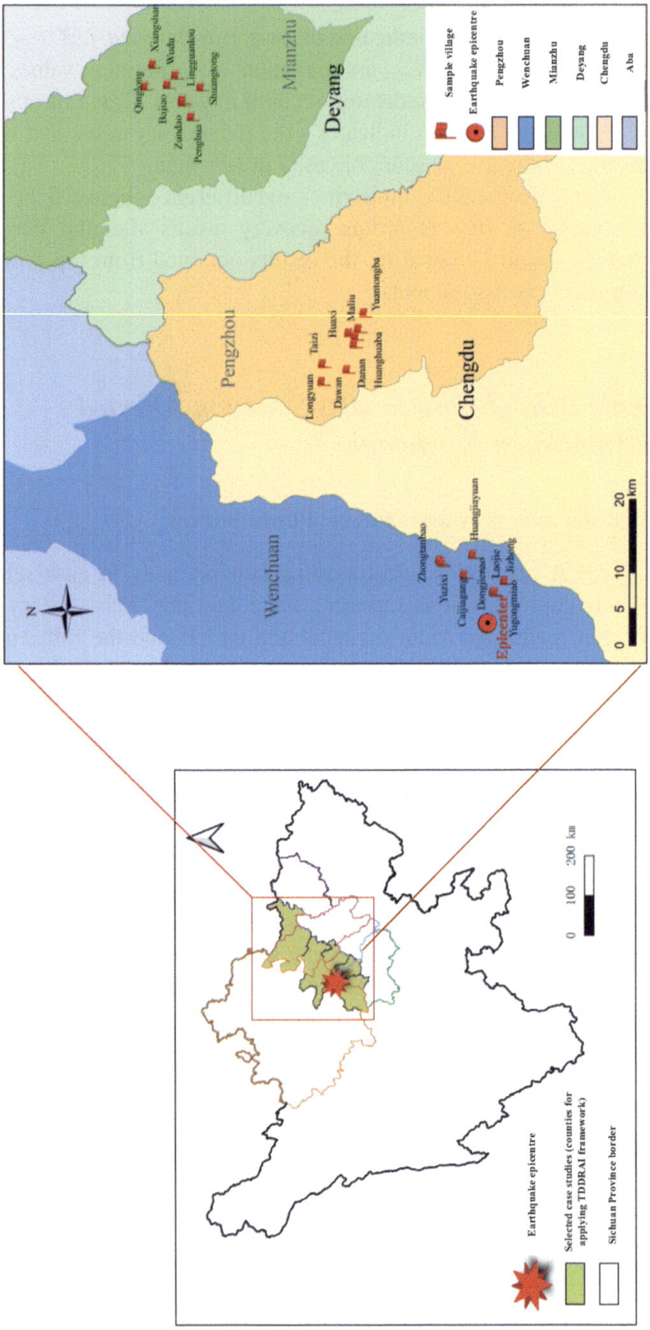

Fig. 7.3 Selected towns/villages/towns for bottom-up disaster recovery assessment- Wenchuan earthquake

Table. 7.1 The selected counties/towns- Data from, adopted from (Chen & Booth, 2011)

Cities/Counties	Earthquake intensity	Deaths (people)	Number of collapsed houses (units)	Counterpart assistant
Wenchuan	VIII–XI	23,871	608,198	Guandong Province
Beichuan	IX–XI	20,047	347,856	Shandong Province
Mianzhu	IX–X	11,380	1,397,925	Jiangsu Province
Shifang	VIII–X	6,132	1,006,921	Beijing
Qingchuan	VIII–X	4,819	714,084	Zhejiang Province
Mao county	VIII–X	4,088	300,229	Shaanzi Province
An county	VIII–X	3,295	774,896	Liaoning Province
Dujiangyan	IX–XI	3,388	655,265	Shanghai
Pingwu	VIII–X	6,565	299,557	Hebei Province
Pengzhou	VIII–X	1,131	622,066	Fujian Province

pioneering way to mobilize the national resources to help the affected areas effectively (Xu & Lu, 2013). The Chinese central government assigned 19 assistance provinces (cities) to assist the 24 counterpart counties in three provinces that have suffered the most. The counterpart assistance focus was not limited to financial aid; the supporter provinces were also required to help their assigned county in architectural design and engineering consultation, repairing or reconstruction of houses, public services facilities and infrastructures, providing construction materials and medical equipment, and encouraging the enterprises in their province to invest in that hit county (Huang et al., 2011; Lingling et al., 2008). Therefore, given each area's difference in recovery assistance for field works in Sichuan province, this study selected ten hardest-hit counties with dissimilar counterpart assistance partners for top-down recovery assessment, as presented in Table. 7.1.

7.3 Cross-Country Comparison of Recovery Assessment Approaches

Cross-country comparisons of disaster recovery assessments can be further divided into top-down and bottom-up recovery assessments.

7.3.1 Top-Down Recovery Assessment

The macro-level assessment of disaster recovery evaluation dimensions after the Bam and Wenchuan earthquakes is given by cross-comparison of the recovery results considering economic, built-environment, and social recovery. Later, the research applies SWOT analysis for each case to assess internal and external conditions that negatively and positively affected each case's recovery process, as presented in Fig. 7.4.

- **Economic recovery**: Despite the initial financial support from the World Bank, the Bam recovery was majorly supplied by the government due to the reduced amount of the World Bank loan caused by the reported shortcomings in the project management and difficulty gaining international loans due to the USA sanctions (Arefian Farnaz, 2018). On the contrary, China has benefited from a broader range of internal budget resources, such as the financial market, central government, revenue from suffered provinces and other provinces, donations, counterpart assistance, and international assistance (Sichuan Provincial Department of

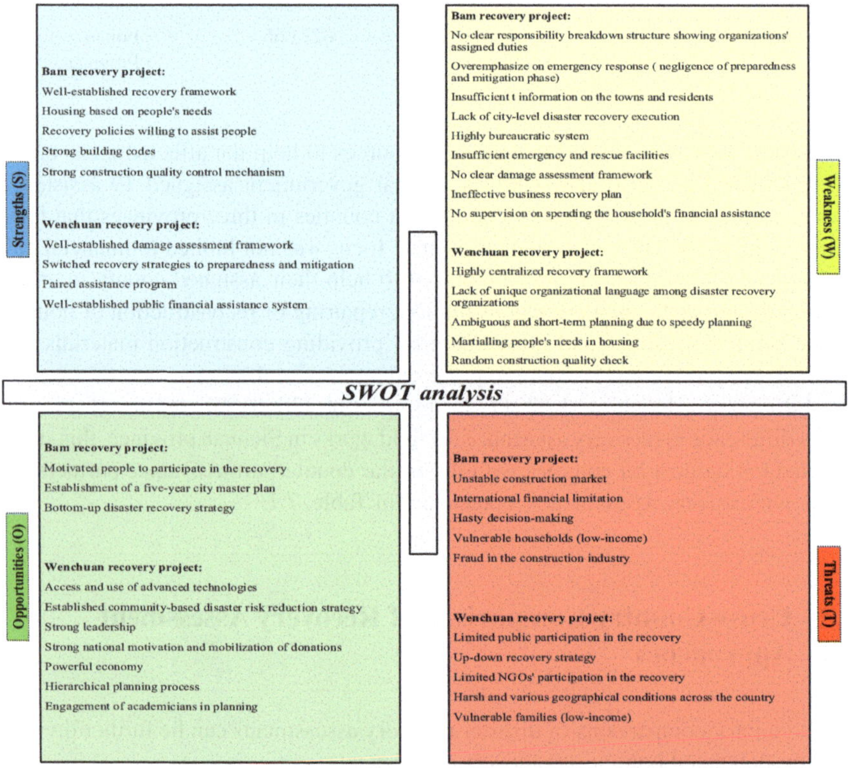

Fig. 7.4 SWOT analysis of the Bam and Wenchuan disaster recovery projects

Housing & Construction, 2013). The reconstruction of the houses after the Bam earthquake was supported by grants, banking loans, and free technical services (Arefian Farnaz, 2018). The loans were directly paid to the households' bank accounts depending on the progress of the construction (Arefian Farnaz, 2018). As the most severe parts were located in urban areas, the higher rate of grants and loans were paid to the urban residents, with the lower interest rates for loans being assigned to rural residents (The World Bank, 2004). The Wenchuan earthquake applied a relatively more complex financial assistance system. Housing financial assistance varied for urban and rural residents, depending on the damage level, financial status before the earthquake, and the household's size. The farmers have been asked to either live in the prefabricated houses built by the government or receive US$ 300 to build their transitional dwellings (Lingling et al., 2008; H. Zhang et al., 2018). For permanent housing reconstruction in Sichuan rural areas, on average, US$ 3,000 has been allocated to each household (Lingling et al., 2008). The priority has been given to low-income families (Ministry of Natural Resources of the People's Republic of China, 2008). Moreover, the Chinese government has carefully designed life support assistance packages for different survivors.

The business recovery for the case of Bam significantly lacked utilizing a well-established detailed program to address all the survivors' livelihoods for the long term (Arefian Farnaz, 2018; Fallahi & Arzhangi, 2014). After the Bam earthquake, people's careers mostly changed due to unemployment, failure to revitalize the old business, smaller business units, and the need to buy a new business place. The business owners believed that the fund amount and allocation by the government were unfair, considering different business units' demands and status (Fallahi & Arzhangi, 2014). In China, taking advantage of the culture of the minority groups and the fertilized land, the suffered areas' business has switched to tourism and eco-agriculture. Considering the government's clear guidelines for business recovery, 93 per cent of the affected enterprises and commercial units resumed operation within three months after the earthquake. However, the shutdown of small shops, changing their function, and dropping the tourism level have widely happened in hard-hit regions years after the earthquake (Wang et al., 2013).

- **Built-environment recovery**: Despite the considerable destruction volume, Bam housing reconstruction was an in-situ project. In China, the general principle for resettling the disaster-affected population was to conduct the in-situ reconstruction or nearby resettlement within the planned areas instead of organizing large-scale emigration (Sichuan Provincial Department of Housing & Construction, 2013). In Iran, the housing reconstruction attempted to consider people's needs and culture by providing 520 design portfolios from the selected design companies (Arefian Farnaz, 2018). The government's national advertisement selected the housing contractors two months after the Bam. However, due to difficulties obtaining housing permission, the chosen contractors left Bam, leaving opportunities for small contractors who took the first loan instalment and escaped while the house construction was not finished. Later, the government decided to let the owners become their houses' contractors and then provided technical supervisors to check the quality of the construction. The quality control of the structures and building code enforcement in Bam has passed several layers (Arefian Farnaz, 2018). On the other hand, in the Wenchuan recovery project, the designs made by unfamiliar designers failed to meet the people's needs, and the culture did not satisfy them. Villagers mostly complained about the ignorance of their lifestyle in the designed dwellings (Xiao et al., 2020). The housing reconstruction methods depended on the location of the damaged house, the residents' financial status, and the residents' willingness to participate in the reconstruction. In the rural areas, self-construed houses stood first, followed by reinforced and government-built houses. On the contrary, reinforced houses hold the first rank in urban areas, testifying to minor damages to urban dwellings. Self-constructed houses again show a higher rate than government-built houses (Zhang et al., 2018). Although the structures' safety has generally improved due to the establishment and enforcement of the seismic and building codes, the quality control of the housing has been supervised randomly in the case of owner-driven housing (Kapucu & Liou, 2014; Lingling et al., 2008). The farmers have tended to apply dangerous construction techniques, such as concrete plates, for roof construction to save money (Lingling et al., 2008).

Based on the report of The World Bank in 2011, Bam City still had an unfinished look due to the hesitation in establishing the city service despite the slight damage to the critical city infrastructures (The World Bank, 2011). The slow progress of the Bam reconstruction caused many people to live in emergency and temporary homes for a long time (Lu & Xu, 2015). Construction was delayed due to the lengthy housing approval procedure due to a lack of human resources in related organisations and internal conflicts between architectural and technical design entities. Later, the increased price of the materials and the householders' financial difficulties postponed the construction phase (Arefian Farnaz, 2018). On the other hand, after the Wenchuan earthquake, where the total losses related to the infrastructure amounted to US$ 28.24 billion (Lingling et al., 2008), the physical reconstruction was accelerated within two years, significantly improving infrastructure (Dunford & Li, 2011; Lu & Xu, 2015; World Bank and Global Facility for Disaster Reduction and Recovery, 2020). The urgency of the reconstruction process has been seen as a two-sided sword as it has had some negative impacts. The insufficient information on the project due to the time pressure has triggered implementation challenges. Moreover, the soft aspects of the recovery have been neglected due to overemphasizing the tangible and physical outcomes (Sichuan Provincial Department of Housing & Construction, 2013; Xiao et al., 2020). Some facilities, such as water connections and water usage fees, did not seem affordable to low-income households (International Federation of Red Cross & Red Crescent Societies, 2011). Years after the Bam earthquake, evacuation facilities and hazard warning systems in Bam did not significantly improve. In contrast, China's disaster monitoring system has improved using a comprehensive satellite-based platform (World Bank and Global Facility for Disaster Reduction and Recovery, 2020). In addition, several evacuation shelters alongside the centres for socio-natural hazard observation, disaster prediction and information, and disaster reduction and mitigation have been established in China (Kapucu & Liou, 2014; National Development and Reform Committee (NDRC), 2008).

- **Social recovery**: After the Bam earthquake, the government of Iran immediately started the search and rescue project. Bam districts were divided into fourteen zones, and fourteen governmental provinces in Iran were chosen to provide help for the rescue operation (Arefian Farnaz, 2018; Fallahi, 2003; The World Bank, 2004). However, due to the lack of experience among the rescuers and the shortage of professional rescue facilities, the rescue was not very successful (Fallahi, 2003; Ghomian et al., 2016). Tents and transitional units (Conex) have been widely used as safe shelters. However, alternative solutions have been considered due to the residents' discomfort living in the government-built transitional units. In Bam, the government decided to establish prefabricated units in the yards of each family, enabling individuals to resume their social networks and privacy quickly. The rescue operation after the Wenchuan earthquake was officially initiated two hours after the quake, when the Chinese Prime minister, Wen Jiabao, flew to the Sichuan earthquake disaster area to lead the rescue teams. However, the shortage of professional rescue equipment and trained rescue forces negatively affected the overall performance of the operation. Moreover, the rescue forces had difficulty reaching some of the hit areas in mountainous regions, causing delayed emergency response (Sichuan Provincial Department of Housing & Construction, 2013). After the Wenchuan earthquake, the prefabricated units in the camps were positioned far from the previous houses and farmlands. Therefore, the government provided extra subsidies for families who did not want to live in the prefabricated units and encouraged them to combine transitional homes with permanent dwellings (Lingling et al., 2008; Sichuan Provincial Department of Housing & Construction, 2013).

The Bam recovery project owned a people participation rate of 138 per cent, much higher than the initial estimation (The World Bank, 2010). Moreover, every 30-60 households selected two representatives (one man and one woman) and established "Town Council's Friend", attempting to share the reconstruction reports through a weekly newspaper (Arefian Farnaz, 2018). The Bam earthquake was a great opportunity to enhance the local people's knowledge about safe construction, as the householders were the supervisors and contractors of their housing project (Arefian Farnaz, 2018). Additionally, some NGOs have provided residents with technical workshops on the structures' earthquake vulnerability, emergency management, and community-based first aid (International Federation of Red Cross and Red Crescent Societies (IFRC), 2004). In China, according to the Asian Development Bank's report, the participation of the affected people, especially women, in the disaster recovery project was not satisfying (Lingling et al., 2008), resulting in a lower level of public disaster risk knowledge after the Wenchuan reconstruction (Ge et al., 2010). Although some households were asked about their needs in housing design, the government has made the final decision (Lu & Xu, 2015). The International Federation of Red Cross and Red Crescent Societies (IFRC) held training sessions for locals, especially schools, teachers, and volunteers, to assist their communities in handling emergencies. These pieces of training were in line to improve the capabilities of community-based health and first aid (International Federation of Red Cross & Red Crescent Societies, 2011).

7.3.2 Bottom-Up Recovery Assessment

In Sichuan province, micro-level disaster recovery assessment was carried out by conducting a public survey in the selected villages, which resulted in the participation of 599 individuals, of which about 53 per cent were females, and 77.12 per cent of the total sample size was over 50 years old. A majority of 69 per cent described the earthquake destruction as complete destruction, where infrastructure was severely damaged with widespread devastation, rendering areas uninhabitable. The financial loss was the most common loss caused by the earthquake, accounting for 57.6 per cent, followed by mental health challenges or emotional difficulties (32.5 per cent) and loss of family members or friends (9.4 per cent). The earthquake caused extensive destruction, leaving only 0.4 per cent of the interviewees unaffected by any losses. Most of this study's population still lives in the reconstructed houses after the earthquake. In Bam, among 122 participants, 46% were female, and almost 42% of the total respondents were between 31 and 40 years old. According to the survey, 61% of them described the earthquake as a severe quake, causing damage in most buildings. Years after the earthquake, most of this population still lived in the reconstructed houses.

The calibrated recovery scores for each sample in every the outcomes across the villages/ districts in each case are given in Fig. 7.5. In Sichuan, the largest difference among the cases is tracked in the economic recovery score between Xiangshan (0.958) and Laojie (0.259). Nonetheless, the recovery of the built environment in Bam exhibited the greatest discrepancy, with Ansari achieving the highest score of 0.54 and Amir-al-momenin attaining the lowest score of 0.28. In Bam, Amir-al- momenin has been identified as the weakest case in all the recovery outcomes. Notably, the primary analyzed data shows that in both Bam and Wenchuan recovery projects, the built environment recovery illustrates the lowest mean value across all villages-districts.

Next, eight models depicting how people perceive the success or failure of recovery were constructed. Table. 7.2 provides a summary of results over three recovery dimensions in Bam and Wenchuan recovery projects. The results revealed that social and economic recovery pathways could be reached through one pathway in the Wenchuan disaster recovery project, contrary to a more complex pattern we previously observed in the Bam recovery project. In Bam, efficiency in emergency response (a_1) and public engagement in recovery project (a_5) were critical conditions forming three pathways of social recovery (Bahmani & Zhang, 2022b). However, in Sichuan, the residents believe that having suitable houses, even if there was an insufficiency in empowering the residents' coping abilities when encountering current and future disasters ($a_3 * \sim a_4$), could effectively lead to successful social

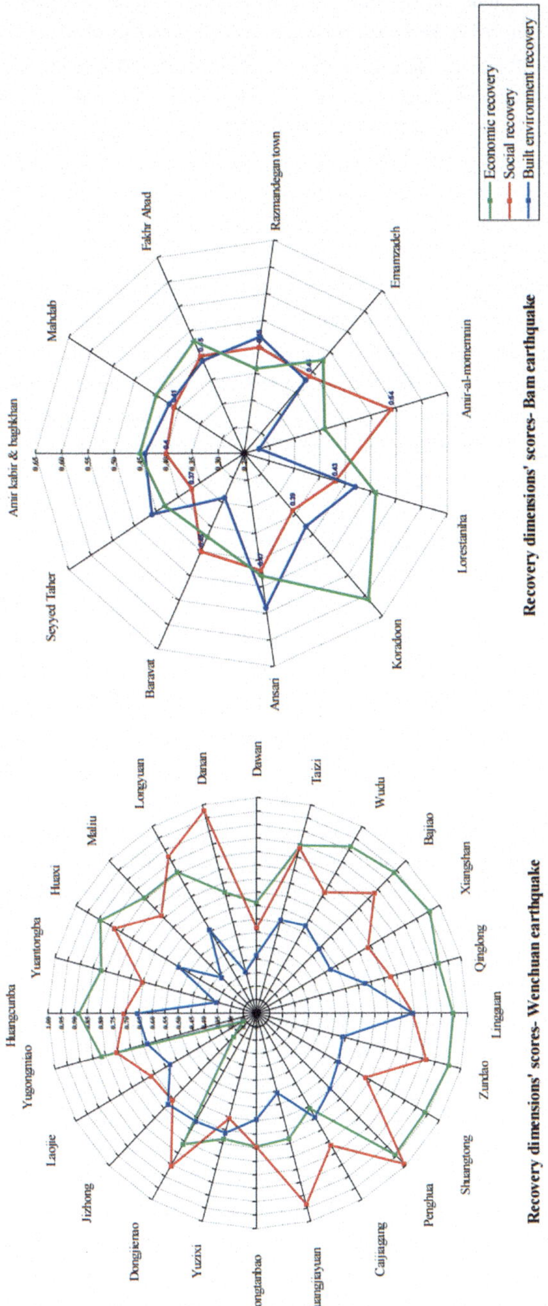

Fig. 7.5 Recovery dimensions scores

and economic recovery. Given the region's pre-earthquake conditions and the instability of local housing, it is plausible that government efforts to provide safe and suitable homes after the Wenchuan earthquake significantly improved living conditions. This focus on rebuilding the built environment increased residents' satisfaction following the reconstruction. The study results also indicated that success in social and economic recovery was closely linked to the Wenchuan recovery project as they have common successful pathways. However, in Bam, economic recovery could be achieved through four pathways, and among the conditions, public participation (a_5) was the most frequently mentioned condition in building the economic success pathway (Bahmani & Zhang, 2022b).

Unlike the more straightforward pathways for social and economic recovery after the Wenchuan earthquake, the failure pathways for these two outcomes were notably more complex than in the Bam case. Sichuan residents believed that social recovery might be seen as a failure if insufficient attention is paid to improving locals' capabilities to cope with disasters and their active participation in housing recovery, even if rescue efforts were well-provided ($a_1 * a_4 * \sim a_5$). Failure in social recovery can also be due to a lack of public engagement in reconstructing their houses, even if they had adequate access to shelters and life-saving facilities during shelter time ($a_2 * \sim a_5$). Meanwhile, economic recovery failure resulted from simultaneous negligence in enhancing public disaster coping capabilities and ensuring their participation in ongoing recovery efforts ($\sim a_4 * \sim a_5$). This finding implies that social and economic recovery are the results of long-term recovery efforts. Even though lack of public engagement ($\sim a_5$) has owned the lowest necessity score in the Wenchuan recovery project's social and economic failure pathways, it appeared in all the failure pathways. Bearing in mind that in the Wenchuan recovery project, public participation (a_5) was not observed in social and economic success recovery pathways, one can conclude that housing was emphasized in forming successful economic and social recovery. However, limited public engagement in recovery programs could cause their dissatisfaction with the social and economic recovery results. This aligns with previous findings from southeast China, highlighting the importance of public involvement in successful recovery efforts (Zheng & Chou, 2024).

In Wenchuan, people considered the rehabilitation of the built environment successful if they were provided with suitable housing that met their needs and if they were highly involved in the reconstruction process ($a_3 * a_5$). Our analysis also revealed that even if the rescue facilities were not highly satisfactory, providing the residents with well-designed houses could also lead to successful recovery from the built environment ($a_3 * \sim a_1$). Therefore, housing and its condition were the most often mentioned factors in successful pathways in the built environment. Meanwhile, in Bam, public participation (a_5) appeared in three of four built environment success pathways and was a condition where Bam residents ranked it highly critical for the success of rehabilitation of the built environment (Bahmani & Zhang, 2022b). This study also proved that, based on Sichuan residents' perception, an insufficient level of people's engagement in housing reconstruction, along with a deficiency in improving people's coping abilities, would lead to failure in achieving successful built environment recovery ($\sim a_4 * \sim a_5$), where the former was the only necessary

Table. 7.2 Comparison of bottom-up recovery assessment scores in Bam and Wenhcuan earthquakes

		Bam earthquake			Wenchuan earthquake		
Outcomes		Pathways	Coverage	Consistency	Pathways	Coverage	Consistency
Social recovery	Success path(s)	a1 * a5 * ~ a2 a1 * a2 * a4 * a3 a2 * a5 * ~ a4	0.93	0.85	a3 * ~ a4	0.88	0.967
	Failure path(s)	a1 * a2 * ~ a3	0.79	0.94	a1 * ~ a4 * ~ a5 a2 * ~ a5	0.803	0.602
Built environment recovery	Success path(s)	a5 * a2 * a4 a5 * a2 * ~ a1 a1 * a3 * a2 * a4 a1 * a5 * ~ a2	0.99	0.84	a3 * ~ a1 a3 * a5	0.976	0.71
	Failure path(s)	a1 * a2 * ~ a3 * ~ a5	0.71	1	~ a4 * ~ a5	0.69	0.871
Economy recovery	Success path(s)	a1 * a5 * ~ a2 a1 * a5 * a4 a2 * a1 * a3 * a4 a2 * a5 * ~ a3 * ~ a1	0.99	0.84	a3 * ~ a4	0.836	0.966
	Failure path(s)	a1 * a2 * ~ a3 * ~ a5	0.71	0.98	~ a4 * ~ a5	0.806	0.520

*** The sign ~ indicates the absence of that condition. The sign * indicates AND, while the relationship between each pathway (of one outcome) is OR

condition. In contrast, the Bam case presented a combination of several conditions resulting in unsuccessful built environment recovery.

Next, an equal-weighted average of three recovery dimensions was calculated to assess the success and failure pathways of integrated recovery for the Wenchuan project and the results are presented in Figs. 7.6 and 7.7. The integrated recovery was seen as successful if social and economic recovery could achieve a high satisfaction level among the residents (social recovery * economic recovery). In this regard, the negation of the success recovery pathway model showed that well-reconstructed infrastructures could not compensate for the failure to provide job opportunities and a steady business environment to call those villages successful examples of integrated recovery after the earthquake (built environment recovery * ~ economic recovery). In addition, in case of failure in built environment recovery, even if social recovery gained a highly satisfactory score, the integrated recovery efforts in those villages are far from successful (~built environment recovery * social recovery). These findings indicate that while the built environment recovery is essential, social and economic policies may be the primary levers to focus on to drive successful recovery rather than relying solely on built environment efforts. A comparison between the two selected disaster recovery projects can highlight both the similarities and differences in disaster recovery assessment approaches. While public emphasis on housing has obviously impacted the integrated recovery's success in both cases, public engagement did not play a significant role in forming success pathways in the Wenchuan project. In contrast, in Bam, public involvement was present in two distinct pathways, potentially changing the result of integrated recovery.

In conclusion, the recovery processes in Bam and Wenchuan highlight key differences in both top-down and bottom-up recovery approaches. While the Bam recovery project faced challenges with financial and housing recovery, the Wenchuan project benefited from more comprehensive government support, although social and economic recovery in both regions showed varying success pathways.

7.4 Comparison of Public and Government Perspectives on Recovery Outcomes: A Closer Look at the Wenchuan Earthquake

This section compares the top-down and bottom-up recovery assessment methods by closely examining the trends in Wenchuan's recovery dimensions over the reconstruction years and comparing them with data collected from administrative sources. In pursuit of this goal, in Sect. 7.4.1, the TDDRAI framework has been used, and after collecting the relevant data over the years, entropy weights for recovery dimensions were calculated. The study then intended to more effectively compare the outcomes of recovery efforts with people's opinions regarding these efforts. Then, in Sect. 7.4.2, we used sigmoid normalization to convert the data produced by entropy scores and fsQCA into a consistent 0 to 1 range and compared top-down and bottom-up recovery assessment scores in three counties.

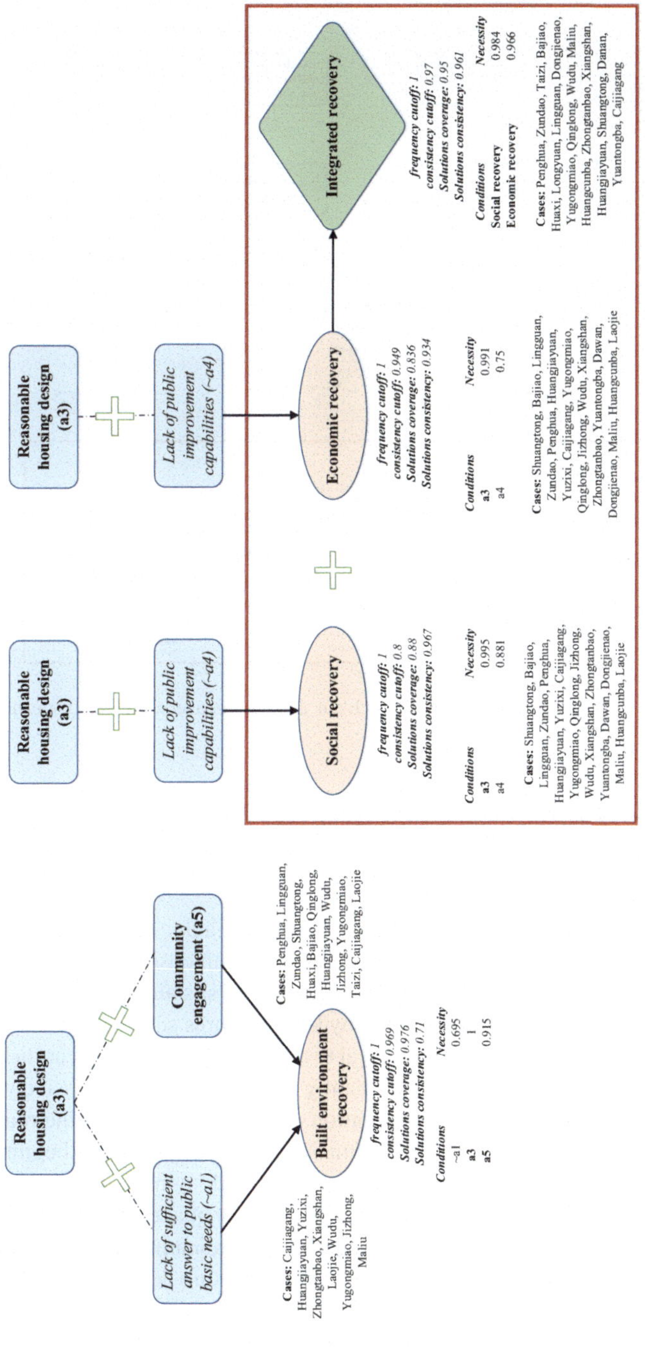

Fig. 7.6 Successful recovery pathways of the Wenchuan disaster recovery project from the public perspective

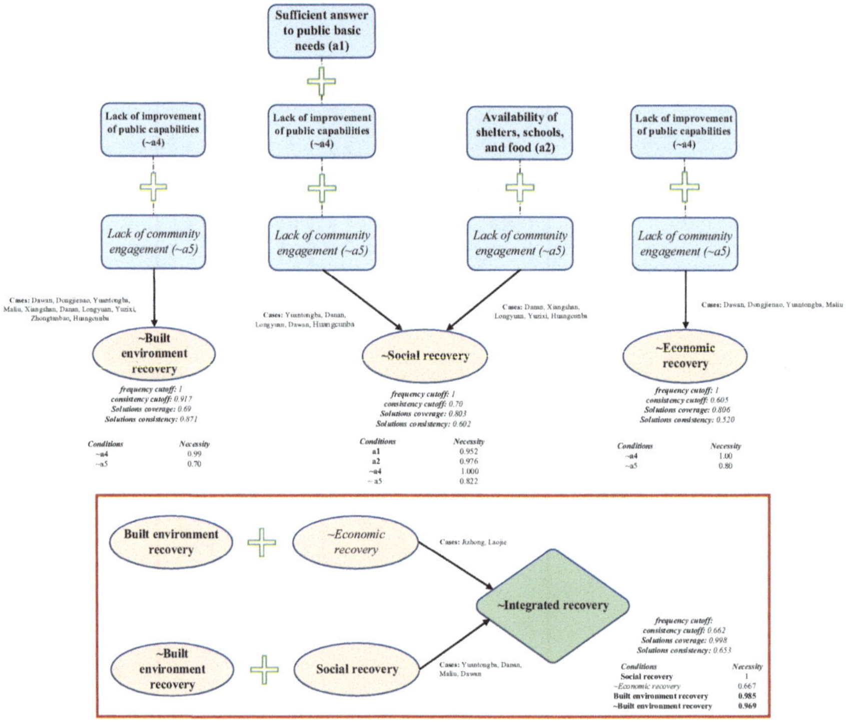

Fig. 7.7 Failure recovery pathways of the Wenchuan disaster recovery project from the public perspective

7.4.1 TDDRAI Framework and Recovery Trends Across Counties

As determined by the weights of the indicators given in Table. 7.3, Sichuan's chosen region saw a robust recovery across several recovery dimensions in the aftermath of the Wenchuan earthquake. The region's capacity to sustain employment levels despite early interruptions was highlighted by key measures such as consistent rural employment rates and the percentage of rural to total employment. Per capita GDP, a measure of economic recovery, fell significantly in 2009 before progressively rising by 2011, indicating the influence of restoration efforts. The length of the roadways varied slightly between 0.090 and 0.098 km in the following years, beginning at 0.092 km in 2007. This suggests that efforts were made consistently to maintain and perhaps enhance the highway infrastructure after the quake. Investment in fixed assets fluctuated significantly between 2007 and 2011, indicating intense post-quake investment efforts aimed at reconstruction and infrastructure development, followed by a decline in investment activities over time. Healthcare and educational facilities showed resiliency, with minor variations in the number of hospital beds and educational institutions, indicating ongoing improvements. Social welfare services initially faced challenges, as evidenced by fluctuations in the number of welfare homes, but began to recover in subsequent years. These indicators demonstrate Sichuan's proactive response to the Wenchuan earthquake, reinforced by consistent infrastructure development, economic recovery, and social stability following the disaster.

The recovery scores of each case are then geographically depicted in Fig. 7.8 to show the temporal and geographical recovery trend. In 2007, the condition of the built environment was relatively stable in the region. However, most cases in 2008 showed severe infrastructure loss after the earthquake, which immediately impacted the region's built environment. Therefore, in most cases, recovery efforts are mainly focused on built environment recovery. The analysis showed that since 2008, the built environment recovery dimension has steadily improved across most cases. After 2010, the recovery of the built environment could successfully surpass the baseline condition in 2007. However, Wenchuan and Mao counties experienced little change over the years, with their built environments remaining largely intact.

One year prior to the earthquake, the economic baseline condition was relatively more robust in the central and eastern parts of the region. A comparison between 2007

Table. 7.3 Entropy weights for each indicator over the years

	Indicators	2007	2008	2009	2010	2011
Economic	Rural employment rate	0.095	0.093	0.098	0.090	0.095
	Proportion of rural employed to total employed	0.095	0.092	0.100	0.091	0.094
	Per capita GDP (yuan)	0.086	0.084	0.062	0.086	0.091
	Investment in fixed assets (10,000 yuan)	0.071	0.079	0.093	0.083	0.063
B.E	Local healthcare (number of hospital beds) (unit)	0.081	0.079	0.086	0.082	0.084
	Local telephone subscribers (subscribers)	0.083	0.083	0.081	0.084	0.081
	Length of highways (km)	0.092	0.092	0.098	0.090	0.095
	Educational facilities (number of primary and secondary schools) (unit)	0.093	0.090	0.095	0.090	0.092
Social	Population growth rate (per cent)	0.090	0.091	0.092	0.091	0.090
	Student-to-teacher ratio	0.090	0.093	0.100	0.089	0.086
	Number of social welfare homes (units)	0.082	0.085	0.059	0.082	0.088
	Percentage of school-aged children enrolled	0.042	0.038	0.036	0.041	0.043

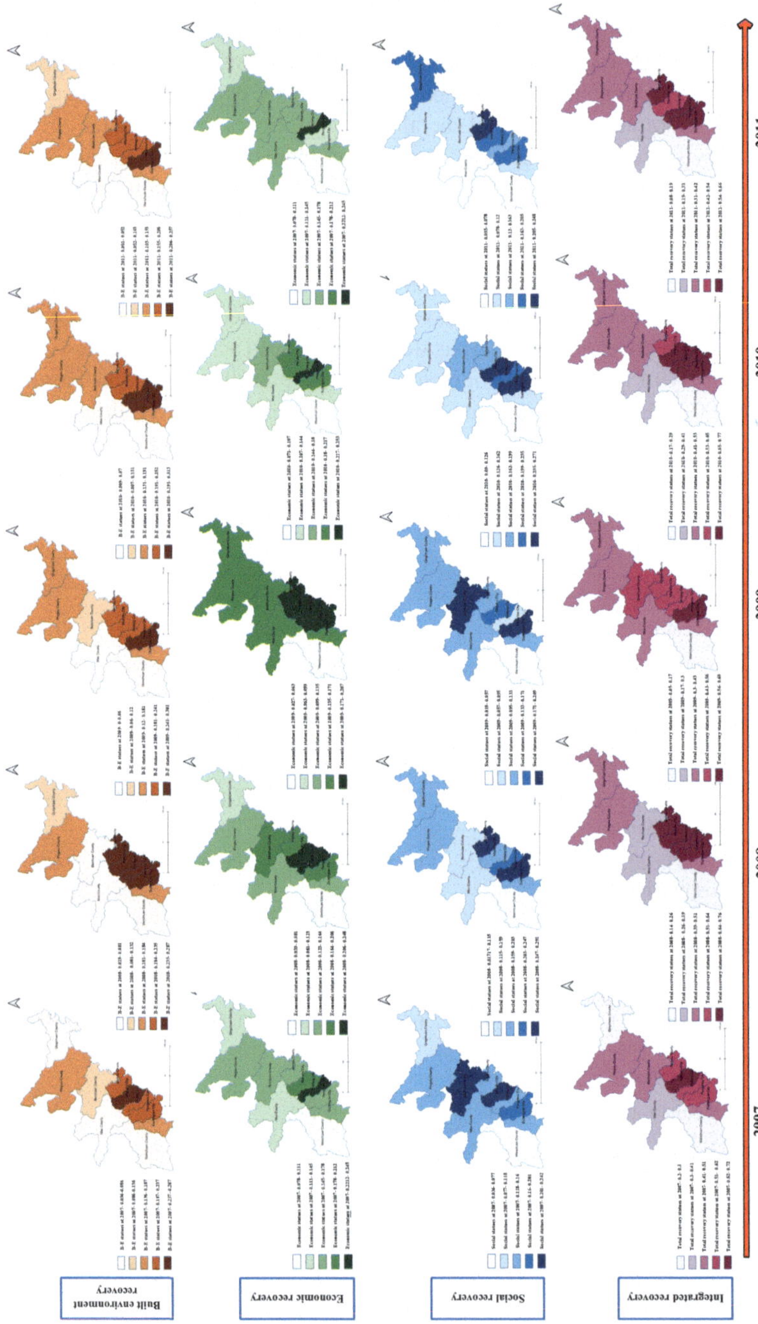

Fig. 7.8 Temporal evolution of disaster recovery after the Sichuan earthquake based on governments' official reports

and 2008 reveals patterns of economic improvement, likely due to ongoing government anti-poverty efforts and the long-term negative impact of the earthquake on the economy. From 2007 to 2009, the economic condition of most regions improved significantly, reaching their most flourishing point. The earthquake's impact on the economy was first observed in 2010, showing a significant decline in eight of ten cases' economic well-being, among which Beichuan and An counties showed the highest decrease in their economic well-being. However, in most cases, 2011 marked the beginning of economic recovery, with conditions returning to normalcy. In some cases, the economic situation improved compared to 2007, stating successful economic revitalization strategies were implemented.

In 2007, the residents' social well-being, except for Wenchuan County, showed acceptable levels. However, the temporal analysis showed that probably because of the earthquake, Beichuan County's social well-being was severely affected, while Pengzhou, An County, Shifang, and Qingchuan experienced rapid social recovery after the earthquake. The recovery of social well-being showed complex patterns. The severe decline in social well-being in 2008 could successfully be compensated in 2009. However, since 2010, social recovery has shown a backward trend, and in many areas, conditions have regressed to below their pre-earthquake status.

The accumulated value of the recovery dimension in each case formed the integrated recovery. The analysis revealed that the Wenchuan earthquake caused a significant decline in the integrated baseline score of all cases. While Pingwu, Dujiangyan, Mianzhu, and Mao counties show some degree of resiliency to withstand the earthquake damages, Beichuan experienced a severe decline in overall recovery. Since 2008, the integrated recovery has been steady and fast as data preserved yearly changes. However, the recovery efforts mostly focused on western areas (Dujiangyan, Pengzhou, Shifang, Miamzhu, and An counties), while other areas could not reach such a high recovery. Notably, although Wenchuan County, the earthquake's epicentre, was the most severely affected area, all recovery dimension scores over the years remained the lowest, and recovery efforts failed to improve the county's overall condition.

The data suggests that the recovery efforts in the Sichuan area after the 2008 Wenchuan earthquake were comprehensive and largely successful in following the principle of "build back better." The steady improvements across the economic, built environment and integrated recovery dimensions indicate that the recovery efforts were well-coordinated and effective in helping the affected communities recover and even surpass their pre-earthquake conditions in many cases.

7.4.2 Recovery Outcome Comparison: Administrative Data vs. Public Perception

The results of the comparison between entropy and fsQCA scores (Fig. 7.9) indicate that in Pengzhou, top-down recovery assessment scores are higher. In contrast, public satisfaction with social and economic recovery outcomes is lower, contrary to what we observed in Mianzhu. Wenchuan, on the other hand, reported similar economic and social recovery performance obtained by both recovery assessment methods, while public satisfaction with built environment restoration stands way higher than the reported process of infrastructure recovery. These differences could be traced back to several underlying reasons. Looking at the region's baseline condition before the quake in 2007, the overall steady condition in Pengzhou might have impacted the residents' required standards for successful disaster recovery, resulting in higher expectations and a perceived lack of recovery success even if minor shortfalls existed. In Miazhu, however, social and economic recovery was efficient based on the public perspective, although top-down assessment showed lower recovery scores. This difference could be because of targeted support in particular areas. A closer look at surveyed data showed that public engagement in recovery efforts was the highest in Mianzhu, possibly contributing to the residents' higher satisfaction with recovery outcomes. In Wenchuan, again, the baseline conditions set lower public expectations of primary living standards, resulting in delighted residents of built environment restoration despite reported shortcomings in this aspect compared to other affected areas. Therefore, the recovery of the built environment of Wenchuan County was more visible to its residents and positively perceived by the public.

In conclusion, comparing top-down and bottom-up recovery assessments in the Wenchuan earthquake case reveals significant differences in public and governmental perspectives on recovery success. While the government focused on infrastructure rebuilding, the public emphasized social and economic recovery, highlighting varying expectations and perceptions of success across different counties.

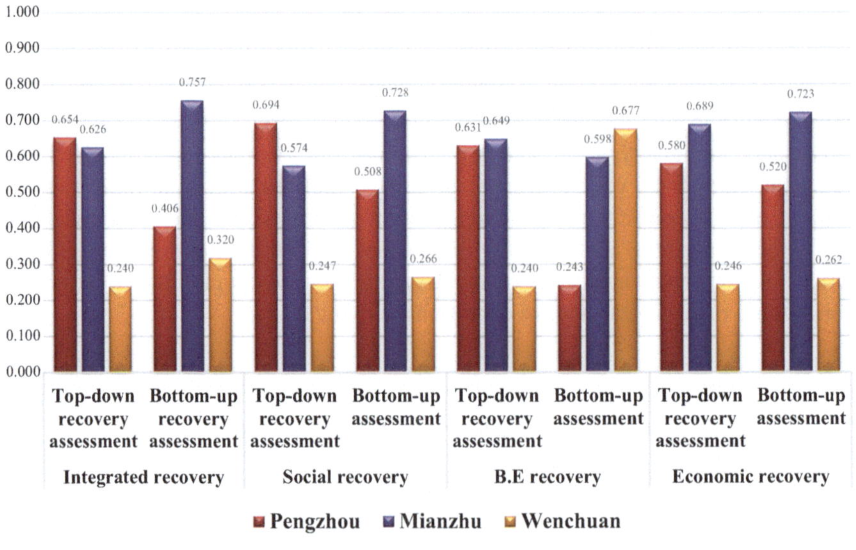

Fig. 7.9 Comparison of top-down and bottom-up recovery scores

7.5 Discussion

This study offers a more realistic understanding of post-disaster interventions by analyzing top-down and bottom-up recovery assessment efforts following the 2003 Bam and 2008 Wenchuan earthquakes. Using the suggested framework by this study would assist disaster recovery managers in bridging this gap and implementing effective recovery strategies that address the needs and concerns of both governments and the public. Based on the research framework design, the findings of this study are categorized into three key areas:

- ***Comparative analysis of top-down recovery efforts in Bam and Wenchuan***: When contrasting the post-quake recovery efforts in Bam and Wenchuan throughout the top-down recovery assessment, apparent differences emerge in the built environment, social factors, and economic aspects. In Bam, the government was primarily responsible for the recovery. Due to a lack of comprehensive planning, the Bam recovery project ended in massive unemployment and little support for long-term business development. Even though the community participation in the project was significant, which helped to convey information about safe construction, the lengthy bureaucratic procedures hindered overall progress. In contrast, Wenchuan benefited from a more sophisticated and varied recovery plan that used various financial sources, enabling quick infrastructure upgrades and more focused financial support. Despite its initial successes, Wenchuan's recovery faced difficulties because of inadequate public involvement, which reduced community resilience and knowledge of disaster risk. Additionally, the built environment in Bam struggled with contractor accountability, while the Wenchuan approach encountered issues with culturally appropriate housing designs. Overall, these comparisons show that although the recovery in Wenchuan showed a quicker and more organized approach, the experiences in Bam highlight how vital community involvement and careful planning are to attaining long-term recovery results.
- ***Comparative analysis of bottom-up recovery assessment in Bam and Wenchuan***: Next, according to the bottom-up recovery assessment, the Bam and Wenchuan disaster recovery initiatives remarkably focused on appropriate housing, but their approaches differed considerably. Following governmental efforts, the Sichuan people reported that having safe and acceptable housing was crucial for social recovery, which improved living conditions and increased life satisfaction. On the other hand, Bam's economic recovery emphasized a more comprehensive strategy that stressed the significance of community involvement. Although less emphasized, the lack of active participation in Wenchuan's recovery routes contributed to perceived failures in social and economic recovery, highlighting this difference in the role of public engagement in the recovery process.
- ***Government vs. public perception of recovery outcomes in Wenchuan***: Significant differences in disaster recovery perception between the government and the public existed after the Wenchuan ***earthquake***, revealing shortcomings in finding common solutions that address both parties' desired disaster recovery outcomes.

The governments focused on the immediate restoration of the built environment due to widespread damage to infrastructures and houses. This emphasis, which continued steadily over the years after the quake, could set a foundation for long-term recovery and aligned with the goal of build back better. Despite this, the social recovery effort has experienced setback trends in most areas. People believe that while rehabilitation of the built environment is crucial, social stability, safety, and business recovery are key drivers for recovery and cannot be replaced by well-planned and reconstructed infrastructures. Therefore, rebuilding the infrastructure and providing the residents with a revitalized built environment should be the fundamental and primary step in paving the path for smoother social and economic recovery. However, the social and economic well-being of the suffered communities should probably be outlined within a long-term strategic plan and monitored to adjust them based on ongoing regional changes. Additionally, engaging communities to comprehend their needs and priorities is another practical solution that can align public and government perspectives on recovery achievements, ensuring greater acceptance of the recovery results by the people as end users. Further analysis illustrated significant discrepancies between top-down and bottom-up assessment scores among the three counties in Sichuan province, which could be traced back to pre-existing conditions in each area, such as the economic strength of the supporting province (Xiao et al., 2020) and lower public expectations of primary county standards. According to existing studies, assigning one supporter province/city to a hit county was unsuitable in most cases since the available resources could be higher than the needs of the suffering county (Lingling et al., 2008; World Bank and Global Facility for Disaster Reduction and Recovery, 2020). Therefore, assigning partnership assistance should greatly match the required assistance level of the suffering area and the economic-technical support power of the assistance partner to avoid imbalanced disaster recovery and dissatisfaction of locals. The pre-existing conditions of the area should also be considered when formulating tailored recovery programs to address recovery requirements successfully and avoid dissimilar recovery patterns across the region.

7.6 Chapter Summary

The comparison between Bam and Wenchuan demonstrates the complexities of disaster recovery. Although both cases had similar recovery aims, the recovery procedures differed dramatically. While the government-led efforts in both cases produced considerable accomplishments, the recovery outcomes were hampered by inadequate public participation in Wenchuan and inadequate planning in Bam. This study emphasizes the significance of integrating both approaches to accomplish full rehabilitation, encompassing physical, social, economic, and cultural issues.

These findings signify the need for having a shared understanding of desired disaster recovery achievements between governments and the public to ensure the recovery is balanced and effective in the entire region. Preferably, comprehensive planning should be given top priority in order to balance short-term infrastructure needs with long-term social and economic development strategies; local communities should be involved in the recovery process in order to build capacity and foster resilience; customized support plans that are in line with local conditions is needed to prevent unbalanced outcomes; government and community cooperation should be encouraged in order to improve understanding and acceptance of recovery initiatives especially in within the project design and planning phases (Davidson et al., 2007); attention should also be paid to culturally appropriate housing designs in order further to support the efficacy and sustainability of recovery efforts.

References

Akiyama, H. (2019). *Who rebuilds?* Civil Society in Post-Disaster Recovery in China and Japan.

Arefian Farnaz, F. (2018). *Organising Post-Disaster Reconstruction Processes Process. Housing reconstruction after the Bam earthquake.* Springer. https://doi.org/10.1007/978-3-319-70911-6

Bahmani, H., & Zhang, W. (2022a). A conceptual framework for integrated management of disasters recovery projects. *Natural Hazards, 0123456789*. https://doi.org/10.1007/s11069-022-05328-5

Bahmani, H., & Zhang, W. (2022b). Why do communities recover differently after socio-natural disasters? Pathways to comprehensive success of recovery projects based on Bam's (Iran) neighborhoods' perspective. *International Journal of Environmental Research and Public Health, 19*(2). https://doi.org/10.3390/ijerph19020678

Befani, B. (2016). *Pathways to change: Evaluating development interventions with Qualitative Comparative Analysis (QCA).* Elanders Sverige AB.

Chen, Y., & Booth, D. C. (2011). The Wenchuan earthquake of 2008 anatomy of a disaster. In *Zhurnal Eksperimental'noi i Teoreticheskoi Fiziki.* Springer. https://doi.org/10.1007/978-3-642-21159-1

Davidson, C. H., Johnson, C., Lizarralde, G., Dikmen, N., & Sliwinski, A. (2007). Truths and myths about community participation in post-disaster housing projects. *Habitat International, 31*(2007), 100–115. https://doi.org/10.1016/j.habitatint.2006.08.003

Devers, K. J., Lallemand, N. C., Burton, R., Kahwati, L., Mccall, N., & Zuckerman, S. (2013). *Using Qualitative Comparative Analysis (QCA) to study patient-centered medical homes* (Issue September).

Di, B., Li, J., Dandoulaki, M., Cruz, A. M., Zhang, R., & Niu, Z. (2020). Household recovery strategies in Longmen Mountain area, Sichuan, China, following the 2008 Wenchuan earthquake disaster. *Natural Hazards, 104*(s1), 123–137. https://doi.org/10.1007/s11069-020-04287-z

Di Gregorio, L. T., & Soares, C. A. P. (2017). Post-disaster housing recovery guidelines for development countries based on experiences in the American continent. *International Journal of Disaster Risk Reduction, 24*(2017), 340–347. https://doi.org/10.1016/j.ijdrr.2017.06.027

Dunford, M., & Li, L. (2011). Earthquake reconstruction in Wenchuan: Assessing the state overall plan and addressing the "forgotten phase." In *Applied Geography* (Vol. 31, Issue 3, pp. 998–1009). https://doi.org/10.1016/j.apgeog.2011.01.001

Fallahi, A. (2003). Review of relief operations and temporary sheltering after the Bam earthquake. *Soffe, 13*(37), 7–23.

Fallahi, A., & Arzhangi, S. (2011). Business continuity after the 2003 Bam Earthquake in Iran. *Communications in Computer and Information Science, 257 CCIS*, 532–540. https://doi.org/10.1007/978-3-642-27207-3_58

Fallahi, A., & Arzhangi, S. (2014). Continuation of small businesses in resistant urban design after Bam earthquake,2003- (In Farsi). *Soffe, 65*, 95–112.

Ganapati, N. E., & Mukherji, A. (2014). Out of Sync: World bank funding for housing recovery, postdisaster planning, and participation. *Natural Hazards Review, 15*(1), 58–73. https://doi.org/10.1061/(asce)nh.1527-6996.0000120

Ge, Y., Gu, Y., & Deng, W. (2010). Evaluating China's National post-disaster plans: The 2008 Wenchuan earthquake's recovery and reconstruction planning. *International Journal of Disaster Risk Science, 1*(2), 17–27. https://doi.org/10.3974/j.issn.2095-0055.2010.02.003

Griffey, A. M. (2010). *Infrastructure disaster management: Insight from small business leaders after a mega disaster* (Issue December). University of Phoenix.

Guarnacci, U. (2012). Governance for sustainable reconstruction after disasters: Lessons from Nias. *Environmental Development, 2*(2012), 73–85. https://doi.org/10.1016/j.envdev.2012.03.010

Guo, J., Bian, Y., Li, M., & Du, J. (2024). Assessing resilience through social networks: A case study of flood disaster management in China. *International Journal of Disaster Risk Reduction, 108*(May), Article 104583. https://doi.org/10.1016/j.ijdrr.2024.104583

Huang, Y., Zhou, L., & Wei, K. (2011). 5.12 Wenchuan earthquake recovery government policies and non-governmental organizations' participation. *Asian Pacific Journal of Social Work, 21*(2), 77–91. https://doi.org/10.1080/21650993.2011.9756108

International Federation of Red Cross and Red Crescent Societies. (2011). *Emergency appeal operation update. China: Sichuan earthquake* (Issue August 2008).

International Federation of Red Cross and Red Crescent Societies (IFRC). (2004). *Iran: Bam earthquake (focus on rehabilitation)* (Issue September-November 2004). International Federation of Red Cross and Red Crescent Societies (IFRC).

Johnson, C. (2006). *Strategic planning for temporary housing: 1999 earthquakes in Turkey.* University of Montreal.

Jordan, E., & Javernick-Will, A. (2013). Indicators of community recovery: Content analysis and Delphi approach. *Natural Hazards Review, 14*(1), 21–28. https://doi.org/10.1061/(asce)nh.1527-6996.0000087

Kapucu, N., & Liou, K. T. (Eds.). (2014). *Disaster and development; Examining global issues and cases.* Springer International Publishing. https://doi.org/10.1007/978-3-319-04468-2

Kermanshachi, S., Bergstrand, K., & Rouhanizadeh, B. (2019). *Identifying, weighting and* causality *modeling of social and economic barriers to rapid infrastructure recovery from natural disasters: A study of Hurricanes Harvey, Irma and Maria.*

Lingling, J., Jiexiu, W., & Lianyou, L. (2008). Technical assistance consultant's report; providing emergency response to Wenchuan earthquake. In *ADB TA7081-PRC* (Issue December).

Liu, Z., & Han, Y. (2009). Post-disaster reconstruction and social work development in the Chinese mainland. *China Journal of Social Work, 2*(3), 221–223. https://doi.org/10.1080/17525090903211796

Lu, Y., & Xu, J. (2015). Comparative study on the key issues of postearthquake recovery and reconstruction planning: Lessons from the United States, Japan, Iran, and China'. *Natural Hazards Review, 16*(3), 04014033. https://doi.org/10.1061/(asce)nh.1527-6996.0000172

Ministry of Natural Resources of the People's Republic of China. (2008). *Policy* interpretation*: A detailed understanding of the urban housing reconstruction policy in Sichuan after the Wenchuan earthquake (in Chinese).*

National Development and Reform Committee (NDRC). (2008). *The overall planning for Post-Wenchuan earthquake restoration and reconstruction.*

Omidvar, B., Abbasi, A. R., Aghajani, M., & Zafari, H. (2016). Evaluation of the residents' satisfaction of reconstruction in Baravat after the Bam earthquake (in Farsi). *House and Rural Environment, 158*(2), 104–116.

Opdyke, A. (2017). *Resilient and sustainable infrastructure systems: A comparative analysis of post-disaster shelter coordination, stakeholder participation, and training.* ProQuest Dissertations and Theses, University of Colorado Boulder. https://doi.org/10.1108/ijdrbe-10-2017-0060

Poudel, N., & Shaw, R. (2024). Challenges in urban-rural food supply chains for disaster resilience in Nepal. *Regional Science Policy and Practice, 16*(8), Article 100073. https://doi.org/10.1016/j.rspp.2024.100073

Sichuan Provincial Department of Housing and Construction. (2013). *5.12 Urban and rural* planning *practice of post-disaster reconstruction in Sichuan (in Chinese)* (Y. Hongbo, Ed.; 3rd ed.). China Construction Industry Press.

The World Bank. (2004). *Technical annex for proposed loan of US$220 million to The Islamic Republic of Iran for Bam earthquake emergency reconstruction project* (Vols. T7637-IR).

The World Bank. (2010). *Implementation completion and results report (loan no.: ibrd-47550) on a loan in the amount of us$220 million to the Islamic Republic of Iran for the Bam earthquake emergency reconstruction project* (Vol. ICR0000112).

The World Bank. (2011). *Bam earthquake emergency reconstruction project.*

Wang, G. Z., Li, Y. M., & Liu, L. P. (2013). Relief effort and reconstruction after Wenchuan Ms8.0 earthquake. *Applied Mechanics and Materials, 275–277*, 2727–2734. https://doi.org/10.4028/www.scientific.net/AMM.275-277.2727

World Bank and Global Facility for Disaster Reduction and Recovery. (2020). *Learning from experience insights from China's progress in disaster risk management.*

Xiao, Y., Olshansky, R., Zhang, Y., Johnson, L. A., & Song, Y. (2020). Financing rapid community reconstruction after catastrophic disaster: Lessons from the 2008 Wenchuan earthquake in China. *Natural Hazards, 104*(s1), 5–30. https://doi.org/10.1007/s11069-019-03789-9

Xu, J., & Lu, Y. (2013). A comparative study on the national counterpart aid model for post-disaster recovery and reconstruction: 2008 Wenchuan earthquake as a case. *Disaster Prevention and Management: An International Journal, 22*(1), 75–93. https://doi.org/10.1108/09653561311301998

Zhang, H., Christophersen, M., Dalen, K., Liu, J., & Pedersen, J. (2018). *Reconstructing a future: Ten years after the Wenchuan earthquake.*

Zheng, Z. Q., & Chou, R. J. (2024). Rebuilding the resilience of mountainous rural communities by enhancing community capital through industrial transformation: A case study from rural Fujian, China. *Habitat International, 149*(February 2023), 103086. https://doi.org/10.1016/j.habitatint.2024.103086

Ghomian, Z., Zade, A. K., Jahangiri, K., Nezhad, S. Y. (2016). Bam earthquake experience. The sudden collapse of life in a few seconds; A qualitative study- (In Farsi). *Journal of Qualitative Research in Health Sciences, 6*(4), 414–426.

Chapter 8
Synthesis and Outlook: Contributions, Challenges, and Future Research Avenues

Abstract This study attempted to address why recovery projects following socio-natural disasters may gain different results. To the date of this study, the related literature suffers from ambiguous definitions of recovery projects' success involving various stakeholders, which tend to be evaluated through different assessment criteria. Therefore, a clear definition of "success" is crucial for these projects. On the other hand, identifying the reasons affecting the dissimilar outcomes in recovery projects can provide us with a list of influential critical factors that affect recovery and assist disaster recovery managers in successfully managing recovery projects through wisely established guidelines. Although the case studies in this book are two destructive earthquakes that happened in Asia, the suggested management framework and recovery success dimensions and critical success factors identification were conducted based on the critical literature review of several resources, which discussed a various range of socio-natural disasters (Table 2.2). Moreover, the Delphi panel that verified those identified factors has global experience dealing with different disasters. However, attention should be paid to applying the results of this study to other cases, and it is necessary to carefully adjust the factors based on each case's specific features.

Keywords Contribution · Research suggestions · Call-to-action

8.1 Contribution to Theory

The first chapter of this book delves into manifesting the needs and significance of this research. The six identified research gaps in Chapter 1 formed the research questions, and a detailed roadmap to proceeding with their answer was proposed in Chapter 2. Applying system thinking and a critical review of the research literature resulted in a conceptual management framework presented in Chapter 3 to identify the most critical disaster recovery steps and suggest appropriate times for their conduction. Chapter 4 developed Primary Success Factors (PSFs) and Selected Success Factors (SSFs) of socio-natural disaster recovery projects.

H. Bahmani et al., *Socio-Natural Disaster Recovery Projects*, Resilient and Friendly Living Environments, https://doi.org/10.1007/978-981-95-0113-7_8

Chapter 4 applied system thinking and factors' interrelationship analysis to place and refine the identified elements in Chapter 3. This chapter's main finding was the objective tree of SNDR projects, which aimed to consider them as ongoing projects. The identified PSFs were then verified through two rounds of the Delphi survey and formed SSFs. Moreover, comparing the standardized results obtained from the Delphi survey and critical literature review revealed the different major points that did not gain enough attention in the literature. Finally, the results obtained from the last chapter were simultaneously compared using the triangulation methodology in Chapter 5. The interaction model between the KPIs and CSFs, which benefitted from applying the logic of Qualitative Comparative Analysis, structured the research's findings based on the prioritization assigned during triangulation and presented in Chapter 5.

In order to investigate the proposed KPIs and CSFs further, the research applied two methodologies. A cross-comparison of the disaster recovery projects that occurred in Iran and China was conducted through Narrative Qualitative Analysis in Chapter 6. The results pointed to the impact of disaster recovery policies on the recovery process. The authors came up with several suggestions and specified strategic recommendations applicable to those two countries and others. The model was adjusted to appraise how the interaction of the distinguished elements might lead to successful recovery per the community neighbourhoods. Chapter 7 significantly appraised successful and unsuccessful recovery solutions by conducting public semi-structured surveys in Bam and Baravat. The recovery was accessed from the perspective of households, as they are one of the most critical stakeholders in SNDR projects. Each chapter's main findings and contributions, alongside the research questions, are demonstrated in Table 8.1.

8.2 Contribution to Practice

This book's findings can also be useful to managers, decision-makers, and policy-makers. Those contributions are listed for each innovation point as follows:

1. Engagement of several stakeholders with various requirements, time pressure, and unidentified project closure time are bottlenecks causing conflicts in SNDR projects. The proposed objective tree of SNDR projects in Chapter 4 can be applied to guide recovery decision-makers and managers to recognize, prioritize, and follow up on recovery goals. This objective tree covering multidimensional subjects can expand the recovery managers' perspective to optimize the resources and efforts based on the prioritized goals. Moreover, it may give the managers a clear idea of the closure time of each stage and smoothen the resource mobilization for the next stage. The management framework in Chapter 3 also facilitates the project management stage in recovery projects.

Table 8.1 Overview of research questions and research contribution

Chapter	Research question	Findings and contribution
3	Which pre- and post-disaster steps should be included in SNDR management?	**Proposing SNDR projects' management framework based on system thinking and the suggested continual life cycle;** • The significance of planning the entire lifespan • The necessity of seeing SNDR projects as a bigger (holistic) system • The necessity of continual stakeholders' participation, communication, updating regulations, and disaster vulnerability reduction
4	Can system thinking assist in identifying the influential factors that influence the success of SNDR projects?	**Identification of PSFs (time-based recovery goals) and SSFs based on system thinking** • Influential interactions between the goals and PSFs • Systematic and holistic organizing of PSFs
4	Which life cycle stages in SNDR projects require further studies to fulfil the real projects' needs?	**Modified PSFs; verified SSFs by the experts' judgment** • Recognition of the most minor and most significant factor based on the Delphi and critical literature review • Identifying the literature shortcomings provided by the comparison of the results • Less focused topics, such as immediate response after a disaster and pre-disaster planning, require further studies based on the experts' judgment
5	How do different factors lead to the successful and comprehensive management of SNDR projects?	**Recognition of KPIs and CSFs for SNDR projects based on triangulation and QCA** • Interaction between the projects' goals and the influential factors • Simultaneous consideration of different possible stakeholders

(continued)

Table 8.1 (continued)

Chapter	Research question	Findings and contribution
6	What combinations of different factors might contribute to successful/failed recovery per community perspective?	**Long-lasting community recovery under the interaction of multiple factors** • The role of public participation and prompt emergency response in successful recovery • The necessity of providing short, mid, and sufficient long-term answers to the project needs • The necessity of appropriate housing design (Reasonable housing design is necessary for recovery success, while unreasonable design is a sufficient and necessary reason for unsuccessful recovery)
7	Can lessons from the previous recovery projects be helpful for recovery decision-makers?	**1. Comparative Analysis of Bottom-Up Recovery Efforts in Bam and Wenchuan** • Housing was a major priority in both cases, but recovery approaches differed: • Wenchuan: The Public emphasized safe and adequate housing for social recovery • Bam: Economic recovery was more community-driven, stressing local involvement • Limited public participation in Wenchuan led to perceived social and economic recovery failures • Bam's stronger community engagement contributed to a more integrated recovery experience **2. Government vs Public Perception of Recovery Outcomes in Wenchuan** ***Government Focus:*** • Prioritized rebuilding infrastructure to "build back better" • Maintained long-term efforts to restore damaged buildings and roads • Assumed that infrastructure recovery would lead to broader recovery success ***Public Perspective***: • While infrastructure rebuilding was appreciated, social and economic recovery was seen as more critical • People prioritized safety, business revival, and social stability over mere physical reconstruction • The lack of social recovery plans led to dissatisfaction despite well-constructed buildings

2. Although the comparison of the viewpoints of recovery experts and researchers (Chapter 4) has contributed chiefly to theory by suggesting the potential research avenues, the lack of conversation between recovery managers and academic researchers is challenging. Recovery decision-makers should actively incorporate with researchers to exchange ideas and use knowledge and experience to solve current or upcoming disasters simultaneously. This research reveals that short-term recovery activities in the aftermath of a catastrophe require further

study. As witnessed in reality, overemphasis on physical recovery overshadows providing sufficient answers for mid-term needs in recovery projects.

3. The KPIs, CSFs, and their interaction model in Chapter 5 can be applied in recovery projects to provide a uniform and integrated report of what has been gained through the recovery process. Those findings which stay neutral against stakeholder groups can holistically and unbiasedly evaluate recovery outcomes without overshadowing intangible aspects such as social recovery. The proposed success dimensions can also be applied as separate or integrated checklists to identify vital parameters affecting the project's process. This study has assessed the KPIs and CSFs through two different case study patterns:

    a. The evidence gained from Bam's neighbourhoods highly emphasized active public engagement in the recovery process. High public participation under government leadership has mutual benefits for people and governments. Through actively engaging the public, the sense of ownership and trust in government would be improved. On the other hand, the governments would face trouble solving the domino effects of the marginalization of people. Within the specific context of each recovery project, the decision-makers should find appropriate solutions to answer public needs and culture and consider their role in future development plans. Additionally, a balanced answer to public needs should be considered during the recovery life cycle, from emergency response to the development phase. The shortcoming in answering short-term needs such as emergency response and transitional housing can negatively affect people's lives and result in public dissatisfaction even though long-term development is sufficiently considered. Regarding the role of appropriate housing in community recovery, houses have to be designed based on the residents' culture and the area's climate to prevent mitigation, abandoned houses, layout modification, and public dissatisfaction.
    b. Looking at the recovery efforts after the Bam and Wenchuan earthquakes side by side sheds light on the pros and cons of both community-driven and government-led systems. Wenchuan's social and economic recovery was hindered by a lack of public involvement despite the fact that it benefited from various funding sources and the quick restoration of infrastructure. On the other hand, community involvement was crucial in cultivating long-term resilience, as shown by Bam's recovery, which occurred despite bureaucratic delays and contractor accountability concerns. To guarantee inclusive and long-lasting disaster recovery, our results highlight the importance of a well-rounded approach that combines top-down planning with bottom-up engagement. These results call for more public involvement, recovery plans that are specific to each area, and funding that is based on needs rather than wants. This research offers disaster managers a practical framework for designing recovery programs that are more inclusive, efficient, and sustainable by bridging gaps between top-down and bottom-up efforts.

8.3 Innovation Points

1. **Project management framework for continual and comprehensive management of socio-natural disaster recovery projects by use of system-thinking**

Lack of simultaneous consideration of pre- and post-disaster planning, overemphasizing hard recovery, and unintegrated recovery tasks assigned to segregated organizations are some of the primary deficiencies in SNDR projects. Through applying system thinking, the complex nature of SNDR projects has been decomposed into five sub-systems, among which project management owning several groups experience high failure rates. Moreover, the conceptual framework for managing SNDR projects presented in Chapter 3 listed the most critical and necessary steps that should be considered in managing disaster recovery projects. The proposed loop life cycle applied within that framework connects mitigation strategies to the development stage and focuses on the necessity of using lessons learnt. The research's suggested management framework also proposes continuous consideration of some soft recovery aspects, the integration stage, within the entire project life cycle. The new concept draws a broader perspective instead of simplifying these projects into a single "reconstruction" phase and is able to facilitate the recovery projects without negligence of intangible outcomes.

2. **Practical success factors and influential pathways emphasizing the role of multilevel factors in the success of socio-natural disaster recovery projects by use of triangulation and qualitative comparative analysis method**

Although several studies identified the need to evaluate recovery projects, no clear definition of success and its effectual factors has yet been introduced. Use of the triangulation of the results obtained in the previous steps, alongside qualitative comparative analysis, provided a solid foundation to structure the CSFs, KPIs, and their interactions in Chapter 5, which is applicable in different recovery projects and can assist recovery managers in recognizing the project goals and identifying the most influential factors. This procedure may facilitate decision-making by saving time and preventing future conflicts. Additionally, recovery success, which was comprehensively and systematically defined, considers various beneficiaries. Therefore, applying the proposed KPIs and CSFs can answer a wide range of beneficiaries' needs.

3. **Assessment of critical success factors and influential pathways within cases; Comprehensive cross-comparison of recovery projects (Bam and Wenchuan) and detailed analysis of disaster recovery project's success in Bam's neighbourhoods based on people perspective by use of fuzzy set qualitative comparative analysis**

Several studies focused on the necessity of using the recovery lessons learned; therefore, a comprehensive cross-comparison of the recovery projects in Iran and China that are highly exposed to socio-natural disasters provided practical suggestions for disaster recovery decision-makers. Several aspects of those cases have been assessed and resulted in the recognition of common obstacles and suggestions (general findings) and SWOT analysis for each case (specific findings). The recommendations proposed in this book can broaden the perspective in modifying/updating disaster recovery workflows.

Additionally, combinations of combinations have been analysed instead of a separate analysis of CSFs affecting the recovery success from the public perspective (as it is commonly used in the literature). The success and failure pathways illustrate that solely considering separate factors' impact on recovery might cause oversimplification of the interactions among CSFs in the complex nature of SNDR projects. Moreover, the correlation coefficient analysis among the CSFs revealed their relationship pattern. Therefore, the success pathways alongside the CSFs' correlation coefficient model should be considered to better manage SNDR projects.

8.4 Suggestions for Future Research

In constructing fsQCA-based model for assessing public perspective toward recovery outcomes, adding up the conditions not included in the model may lead to different success patterns. Note that some conditions might be put aside since they do not vary across the selected cases. Data collection methods are also suggested to be more diversified. Multiple data sources would enable the research to verify the results and procedure through triangulation. More cases might enhance the results regarding the data collection stage during public interviews. Future research may also consider using the entire model and its components introduced in Chapter 5 if there is enough available data, resources, and time.

The top-down recovery assessment framework adopted by this research was modified based on available data at the county level, which may have left aside some critical attributes. Moreover, the research team focused on small regions with the highest destruction levels due to the vast damage they experienced after the Wenchuan earthquake. Future studies could consider adjusting the recovery assessment measures and research areas based on the availability of data and resources. There is still room to modify the TDDRAI framework to work perfectly for cases with insufficient data.

Future research should focus on overcoming several critical shortcomings in disaster recovery studies. Initially, it is crucial to examine the effect of community participation on recovery outcomes, specifically how various types of public engagement affect both recovery time and community resilience. This may yield significant information regarding the effectiveness of diverse participation strategies in disaster recovery. Moreover, future studies should investigate long-term recovery in rural areas, focussing on how recovery strategies might be customized to address the distinct needs of these locations. The findings of this study can be evaluated further by incorporating them into current disaster recovery frameworks, particularly by highlighting public involvement in recovery efforts. By incorporating these insights, governments can enhance their national recovery strategies, especially when inequality exists between top-down and bottom-up initiatives. Subsequently, the next studies should encompass global comparisons of post-disaster recovery initiatives across many cultural contexts. Such comparisons may facilitate the development of more adaptive, culturally relevant recovery models applicable in many disaster-prone regions. Table 8.2 sheds light on the difficulties of this research and the decisions made by the researchers.

Table 8.2 Overview of research difficulties and subsequent response

Occurrence phase	Difficulty	Response	Pros and Cons
Literature review	A large number of published studies	Keyword-based selection based on a variety of topics Critical selection from the most relevant journals	Provide a wide range variety of resources Choosing relevant topics
Basic evaluation	Shortage of study to render a comprehensive outlook of SNDR projects	System thinking to organize the topics Tripartite measurement tools; Critical Success Factors, barriers and, suggestions	Managing the subjects Better categorization of subjects A clear understanding of SNDR projects
Basic evaluation	Heterogenous factors identified throughout the research	Application of system thinking and factors' interrelationship analysis	Show the relationships Summarizing the factors
Delphi Survey	Difficulty in finding panellists	Search through different resources such as published materials, online social networks, and snowball sampling	Identification of network of experts
Case study	Pandemic and difficulty in collecting primary data	Remote data collection (trained interviewers	Fast and reliable collected data Low level of observation
Case study	Limited time and budget for survey	11 neighborhoods in Bam and Baravat, Iran	Data is sufficient for this study; however, it is suggested to survey a bigger sample size
Case study	References are written in three languages	Online translation for Chinese resources and the authors' translation for Farsi ones	Time-consuming Low efficiency A broader perspective of the cases' context
Model proposal and analysis	Transforming qualitative data into scientific research	fsQCA	Effective generalization of the outcomes Time-consuming
Analysis	High number of factors	Tree-shaped categorization of the effectual factors	Simplified model Misinterpretation of some factors upon the reader's perspective

The manufacturer's authorised representative in the EU is Springer Nature Customer Service Centre GmbH, Europaplatz 3, 69115 Heidelberg, Germany. If you have any concerns regarding our products, please contact ProductSafety@springernature.com

Printed and bound by CPI Group (UK) Ltd, Croydon, CR0 4YY
01/10/2026
02225915-0002